**Systems Engineering Models
of Human–Machine Interaction**

NORTH HOLLAND SERIES IN
SYSTEM SCIENCE
AND ENGINEERING
Andrew P. Sage, *Editor*

Systems Engineering Models of Human–Machine Interaction

Series Volume 6

William B. Rouse
University of Illinois, Urbana, Illinois

NORTH HOLLAND
New York • Oxford

Elsevier North Holland, Inc.
52 Vanderbilt Avenue, New York, New York 10017

Distributors outside the United States and Canada:

Thomond Books
(A Division of Elsevier/North-Holland Scientific Publishers, Ltd.)
P.O. Box 85
Limerick, Ireland

Library of Congress Cataloging in Publication Data

Rouse, William B.
    Systems engineering models of human–machine interaction.
    (North Holland series in system science and engineering; v. 6)

    Bibliography: p.
    Includes index.
    1. Man–machine systems—Mathematical models. 2. Systems
    engineering—Mathematical models. I. Title.
TA167.R68   620.8'2   79-25437
ISBN 0-444-00366-5

Desk Editors Michael Gnat; Michael Cantwell
Design Series
Art rendered by Vantage Art, Inc.
Production Manager Joanne Jay
Compositor Comptype
Printer Haddon Craftsmen

Manufactured in the United States of America

Contents

Contents

Preface

In our technological society, humans are increasingly interacting with machines. The designers of these machines would like to consider this human–machine interaction in the same quantitative manner in which they pursue much of the rest of the design process. To this end, mathematical models of human–machine interaction are needed. A wide variety of such models is presented here.

This book was written with several goals in mind. First of all, one objective was to emphasize the current state of the art rather than provide a thorough historical perspective. Based on this goal, well over 80% of the references cited have publication dates of 1970 or later. However, readers interested in earlier works will find excellent sources on this material noted throughout this book.

A second goal was to provide a treatment of a highly mathematical topic while avoiding calculus, differential equations, Laplace and Fourier transforms, and so on. Thus, the only mathematical prerequisites for this book include basic algebra and probability theory. Although representing all of the models algebraically does result in a few topics (but not too many) being covered rather tersely, it is hoped that the purely algebraic treatment will make the material covered accessible to many more readers. Perhaps those readers who find this book stimulating will go on to study those works that require more extensive mathematical prerequisites.

A third goal was to include basic tutorials on the modeling methodologies of interest and thus avoid requiring the reader to consult other basic sources. Furthermore, along with the tutorials, fairly complete examples of applications are discussed and a breadth of other applications briefly reviewed. The choice of the characteristics noted in this paragraph was motivated by a desire to employ this book as a primary text for a graduate course on human–machine systems that typically includes both engineering and psychology students.

For the most part, this book is based on lecture notes used for this course, which is offered in Industrial Engineering at the University of Illinois, as well as on lecture notes for a short course on modeling human–machine interaction given outside the university. In teaching these courses, I have found that the type of material presented in this book is nicely complemented by having students pursue a series of small design projects in which they have to choose among the various available models, resolve measurement problems, and so on. This approach leads students to realize, from experience, the ways in which models are particularly useful. This end is not served as well by specific exercises in which students primarily learn to manipulate equations. For this reason, a set of such exercises is not included in this book. However, if such material is desired, the numerous texts cited throughout this book are more than adequate sources.

To a great extent, this book is also based on the results of the author's interactions with colleagues at Illinois, elsewhere in the United States, and in other countries as well. I am truly indebted to these individuals, who have contributed greatly to the lines of thought formulated in this book. I am also most grateful to Carolyn Robins for her editorial assistance in preparing the manuscript.

Urbana, Illinois   William B. Rouse

**Systems Engineering Models
of Human–Machine Interaction**

Chapter 1
Introduction

Humans interact with machines in many ways. Many people drive automobiles. Some people repair automobiles. Others fly airplanes, work in nuclear power plants, or work with computers. In fact, it is not unreasonable to claim that humans, in general, are increasingly interacting with machines.

As designers of machines, engineers should be concerned with the way in which humans interact with machines. This issue is of importance for several reasons. First, one wants to ensure that machines are safe to operate. Second, one would like to know how the human's abilities and limitations affect the performance of the overall human–machine system. With this knowledge, the machine can perhaps be designed so as to complement the human's characteristics. Another reason for being concerned with how humans interact with machines involves the problem of job satisfaction. We should like to design machines so that interacting with them is satisfying, or at least certainly not demeaning. Although job satisfaction and safety are not concerns of this book, we mention these topics here in recognition of their importance.

This book is concerned with the performance of human–machine systems. Adopting a typical engineering point of view, our goal is to develop methods of analysis that allow one to *predict* performance. This goal should be contrasted with that of trying to *measure* performance. Were measurement our goal, then we would devote this book to the various considerations surrounding the topic of experimental design.

To predict the performance of a human–machine system, we require some representation of the system that allows us to determine how independent variables affect dependent variables. To represent a human–machine system, we shall have to depict both human and machine behavior in compatible terms. For a variety of obvious reasons, it is appropriate to represent human behavior in machine-like terms, as opposed to vice versa. Further, while a good verbal description of the human's tasks, abilities, and limitations is certainly an essential first step towards developing a representation of human behavior, such a description is usually inadequate in that it will only allow *qualitative* statements about the performance characteristics of the human. Our goal is to provide *quantitative* predictions of human performance.

The engineering approach to obtaining quantitative performance predictions for almost any problem is to develop a mathematical model of that problem in terms of how relevant inputs (independent variables) affect interesting outputs (dependent variables). If this approach is successful, and it is not always successful, then the resulting mathematical model can be quite useful in several ways.

The Purposes of Models

In general, there are four major uses of models. First of all, the modeling process is itself beneficial. Developing a mathematical model requires a very organized and thorough pursuit of all the issues surrounding a problem. This is especially true when one gets to the stage of writing a computer program that incorporates the resulting model. At this point, one often finds that various parameters are undefined or perhaps immeasurable. Once the model is sufficiently defined to allow the computer program to produce predictions of performance, then it is not unusual to find the predictions to be ridiculous when compared to the performance of the real system. In this situation, one has obviously overlooked some crucial aspect of the real system. Then, the modeling process iterates and the model is updated. This iterative process of model development and testing provides many insights with respect to the system being studied. These insights are valuable even if the model's predictions are actually never used.

A second use of models is to provide succinct explanations of data. For example, a study of the learning of motor skills may result in a learning curve of rms (root-mean-squared) error versus time. While this tabulation of error versus time summarizes the results of the study, a much more succinct explanation might be found using a single parameter to characterize learning rate within a mathematical model of learning of motor skills. Thus, models can be used to aggregate various plots and tabulations into a few behaviorally meaningful parameters and, in that way, allow much clearer comparisons among tasks and experiments while also providing much simpler "rules of thumb" for systems designers. We shall return to this point repeatedly throughout this book.

Models are also useful for designing experiments. It is not unusual for a model to have many free parameters. Estimating these parameters can present data collection problems. These problems can be lessened if one tests the model to determine the sensitivity of performance predictions to parameter variations. If the model's predictions are fairly insensitive to some of the parameters, then one can assume "reasonable" values for these parameters and avoid investing the effort necessary for estimating them. Similarly, if one is contemplating an empirical study of human performance, one can use the results of a sensitivity analysis with a model to determine which parameters should be varied in the empirical study. This may sound like an unusual procedure. If one had a good model of human performance in a particular task, then why would one run an empirical study of that task? On the other hand, if one did not have a good model, how could one use a model to determine which parameters have the greatest effect on performance? Quite simply, one can often use an approximate model to obtain a feeling for the sensitivity of performance to various parameters. The value of this approach is dependent on how reasonable the model is regardless of the fact that it has not yet been experimentally validated. Using models in this way is part of the art of engineering.

Finally, models are useful in terms of the quantitative predictions that they produce. While this aspect of model usage is often given too much emphasis, relative to the other purposes of models, quantitative predictions are nevertheless useful in the process of designing systems. For example, it is quite important to know how a pilot and aircraft will interact *before* one builds the aircraft. Another example, which we discuss at some length in Chapters 4 and 7, involves embedding predictive models in computer programs in a manner that allows the computer to "understand" the user.

Most of the discussions throughout this book will focus on models in terms of the performance predictions that they produce. However, we want to emphasize here in our early discussions that one should look at such predictions as only one of the benefits of modeling.

The Modeling Process

The first step in the modeling process involves defining the problem. This includes a statement of the phenomena of interest as well as a choice of performance measures. Thus, for example, one might choose to study the abilities of air traffic controllers to detect potential midair collisions. One might designate probability of detection and time until detection as performance measures. As another example, one might decide to investigate the abilities of maintenance personnel in troubleshooting tasks. Appropriate performance measures for this task would be probability of correct diagnosis, time needed for diagnosis, and cost of diagnosis.

While this process of defining the problem may seem quite straightforward, this is far from true. It is not unusual to realize suddenly that one is working on the wrong problem. Also, it is not uncommon to discover that the performance measures initially chosen are inappropriate. This latter difficulty can frequently be attributed to the fact that performance measures are often only indirect indices of desirable system characteristics. For example, vehicle ride quality is difficult to quantify unless one uses indirect measures such as rms amplitude of vibrations, and so on. Using such indirect measures increases the likelihood of choosing inappropriately. Thus, one should not view problem definition as a necessarily easy portion of the modeling process. Indeed, it is probably the most troublesome step in that it is difficult to obtain training (e.g., by reading books such as this) that will give one the ability to define problems insightfully and appropriately. Experience appears to be the key ingredient in this step of the modeling process.

With the problem defined and candidate performance measures chosen, the next step of the modeling process involves representing the problem. What are the system's inputs? What are its outputs? How do inputs affect outputs? How are the performance measures of interest affected by the system's variables? Representing the problem involves providing answers to these questions.

Up to this point, we have not been concerned with formalizing our conception of human–machine systems. However, some formalization is now necessary if we are to avoid confusing terminology. Figure 1.1 depicts a basic human–machine system. With this figure, we want to make three important clarifications. First of all, unless otherwise specified, the term "system" will refer to the overall combination of human and machine. Second, the term "machine" will refer to everything other than the human. Third, and finally, this figure clearly illustrates the notion that the machine's outputs are the human's inputs and the human's outputs are the machine's inputs.

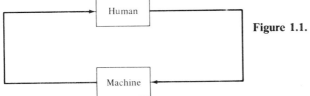

Figure 1.1. Basic human–machine system.

This last idea is central to much of this book. The output of the human's perception/decision/action process may, for example, be the rotation of an automobile's steering wheel. The angle of rotation is the input to the automobile. The machine (i.e., the automobile) converts this angle into a change in the vehicle's path. The human perceives this change and then updates the earlier decision/action regarding the rotation of the steering wheel, and so on.

In terms of representing this problem, one could employ various engineering tools to model how steering wheel inputs affect automobile outputs. However, this book is *not* concerned with this aspect of modeling. Instead, we are concerned with how the path that an automobile is following affects the steering wheel rotations that the human produces. Thus, throughout this book, we shall assume that a model of the machine is available and our discussions will be aimed at developing a model of the human. Of course, as we shall later discuss at some length, the human's behavior is highly influenced by the particular machine involved in the interaction; thus one cannot view the models of human and machine as independent of each other.

Succinctly then, we can view problem representation as the process of developing a relationship between the inputs and outputs of the human. As discussions throughout this book will indicate, the appropriate type of representation depends on a variety of factors. In Chapter 7, we shall consider the problem domains appropriate to the various representation methodologies discussed in this book.

Once a representation of the problem has been formulated, the next step of the modeling process is calculation of performance. Consider the following example. Suppose the input to the human x results in the human producing output $y = a + bx$. Further, assume that the machine transforms its input y into output $x = c + dy$. Finally, the performance measure might be x^2 where the goal is minimization of x^2. Calculation of performance involves determining the simultaneous solution of $y = a + bx$ and $x = c + dy$ and then calculating x^2.

Although this very oversimplified example illustrates the concept of calculating performance, it by no means depicts the usual level of difficulty of such calculations. In fact, it is not unusual to be unable to solve analytically the set of equations embodying the models of human and machine. In such situations, one has to resort to either approximations or simulation.

Simulation solution involves developing an analog and/or digital computer program that emulates the set of equations of interest and allows one to calculate various properties of the resulting solutions. Since human–machine systems problems frequently have probabilistic aspects, this calculation is often of a statistical nature. Thus, one must replicate the simulation

solutions many times to obtain a sufficient number of samples of performance measures to be able to estimate average performance within reasonable confidence limits. Although simulation approaches to calculating per-
formances certainly work, one usually attempts to avoid simulation unless
analytical solutions, perhaps with appropriate approximations, are not
possible. This is simply due to the fact that simulation solutions are
typically more time-consuming and costly than analytical solutions. Nevertheless, as will be discussed at various points throughout later chapters,
many realistic and important problems have to be solved using simulation.

The next step in the modeling process involves experimental validation
of the model. There are two parts to this process. First, since most (but not
all) models typically have several free parameters, usually reflecting some
behavioral assumptions (e.g., reaction time), one may have to experimentally determine the values of these parameters. This involves adjusting the
parameters until the performance of the model matches actual system
performance as closely as possible.

There are many misconceptions about this performance-matching process. Some researchers (i.e., nonmodelers) have claimed that with two or
three free parameters they can make model performance match virtually
any empirical results. This is patently absurd. For example, if the actual
system is such that $y = x^2$ and one chooses to use $y = a + bx$ as a model,
no amount of manipulation of the constants a and b is going to match this
model to the real system over a reasonable range of x. The important point
to note here is that the structure of the model will preclude certain types of
behavior regardless of the values of its parameters. Thus, one should be
careful in deriving a model's structure and, for the most part, not apologize
about having to estimate the model's parameters.

On the other hand, some researchers (i.e., modelers) tend to let parame-
terization get out of hand. Occasionally someone will use 10, 15, or 20 free
parameters to match a single performance measure (e.g., rms error). This
clearly presents methodological problems. However, overparameterization
also presents difficulties in terms of interpreting the meaning of parameter
variations. This subverts some of the purposes of modeling, namely,
providing succinct explanations of data and providing assistance in designing experiments.

The second part of the experimentation process involves using the model
to predict performance and then empirically determining how close the
predictions are to actual occurrences. Of course, if the problem of interest
has probabilistic aspects, then one does not expect a model's predictions to
be perfect. The essential question is this: Does the model provide more
predictive ability than one could obtain by simply using the mean of past
performance as a predictor of future performance? The answer to this
question can be quantified in terms of the percent of the variation about the

mean that is explained by the model. One useful rule of thumb here is the following: The percent of the variance that a model will be able to explain (predict) is inversely related to the robustness of the human–machine system environment being studied. Thus, for example, one can probably model variations of detection time in a pattern recognition task to a greater extent than one can model variations of task time in problem solving situations (Rouse and Rouse 1979).

The results of experimentation lead to the comparison step of the modeling process. Although it is difficult to separate experimentation from comparison, as evidenced by the issues discussed in the last few paragraphs, we nevertheless want to single out some very important aspects of the comparison process. First of all, the distinction between behavior and performance should be noted. Within this book, the term "behavior" will be used to refer to *what* the human does whereas the term "performance" will refer to *how well* it is done. In a tracking task, for example, the human's behavior is the specific time history of control movements produced whereas the performance is the rms tracking error that results.

Since a variety of patterns of behavior might result in the same performance, it is very much easier to develop models to predict performance than it is to develop models to predict behavior. For many engineering applications, performance predictions are all that is necessary. When such is the case, then comparisons of the performance of the model with empirical measurements of performance are sufficient to validate the model. However, such validation does *not* allow one to infer that the model's behavior matches human behavior. If one is interested in behavior as well as performance, then one should make validating comparisons between the behavior of the model and the behavior of the human. Although this principle does not have to be followed religiously, it should not be flagrantly violated.

Since a model that can accurately predict behavior will also be able to accurately predict performance (but not vice versa), a behavioral model is much "stronger" in the sense that it more completely describes the human as he performs the task of interest. This point has been convincingly argued by Gregg and Simon (1967). They contrast process (behavioral) models and statistical (performance) models and defend in detail the notions briefly outlined in these last two paragraphs.

Besides deciding whether comparisons will be based on behavior or performance, one also must choose the range over which validation is to be attempted. Quite often the particular application motivating the modeling effort will dictate the range over which one should vary the conditions for which the model is being tested. One should choose the range of testing conditions so that later predictions will be interpolations rather than extrapolations. If unforeseen situations later cause one to have to use extrapolations, then one is much better off if a behavioral model is available, especially if one is only interested in performance predictions for the extrapolated conditions. The reason for this is obvious. The more fine-grained the validation process is, the more reasonable extrapolations will be, particularly if these extrapolations are not as fine-grained as the validation.

Summarizing our discussions of the modeling process thus far, the following steps of the process are important to note:

1. Definition
2. Representation
3. Calculation
4. Experimentation
5. Comparison
6. Iteration

With the exception of iteration, we have discussed all of these steps of the modeling process. The iteration step has been added at this point to emphasize that the modeling process is really not as straightforward as this introductory chapter may have led one to believe. One typically goes through a seemingly never-ending process of refinement whereby model inadequacies are eliminated and the range of a model validity is extended. Occasionally as one is trying to eliminate inadequacies and extend the range of validity, one finds that a whole new representation methodology is needed. Several transitions of this type will be considered throughout this book. In general, iteration results in the modeling process never terminating, basically because the modeling process is synonymous with the process of knowledge acquisition and organization whose inherent goal is growth and change.

The Use of Analogies

One of the most powerful problem solving methods of science and technology is the use of analogies. Basically, this involves viewing a new problem as if it were an old problem for which one is likely to know the solution or, at least, possess considerable insight. For example, one might view the central nervous system as an electrical circuit and then employ various circuit analysis methods in an attempt to understand the central nervous system. This analogy might prove quite satisfactory until one must deal with the chemical nature of central nervous system activity. Then, the electrical circuit analogy might have to be modified or perhaps replaced.

The iterative process of adopting, modifying, and replacing analogies is central to science and engineering. When an analogy within a particular area of research gathers a sufficient number of adherents, it is often termed a *paradigm* (Kuhn 1962). The emergence of paradigms is a focal point of study for those who research the history of science and engineering. One especially important notion that can be gleaned from this research is that the emergence, and especially the maintenance, of a paradigm can usually partially be attributed to a social consensus among researchers rather than to purely technical considerations (Ziman 1968). This phenomenon results in tremendous inertia in the sense that the need for a new paradigm often has to be overwhelming before an old paradigm is replaced.

This book considers a variety of analogies that are of use when analyzing human–machine systems. Of all the analogies that we will consider, perhaps the only one that has achieved the status of a paradigm is the servomechanism analogy. The basic idea is that the human acts as an error-nulling device (e.g., a thermostat) when driving an automobile, flying an airplane, or doing just about anything. Because of Norbert Wiener's efforts in this area (Wiener 1948), the servomechanism analogy is often referred to as the cybernetic paradigm. However, as we shall discuss in Chapter 7, it is perhaps unfortunate that cybernetics often is viewed so narrowly.

The servomechanism analogy or paradigm has proven to be quite useful because of the great number of control theory methods available for analysis of systems that can be represented as servomechanisms. Chapter 3, in particular, is devoted to discussing some of these methods. However, while the servomechanism paradigm is still predominant in the human–machine systems area, several new analogies have emerged that are also quite useful.

The need for new analogies has been precipitated by the increasing use of automation. The increasing use of computers to perform control tasks has resulted in the human's role becoming more like that of a monitor and supervisor. In such a role, the human can have responsibility for more tasks. Furthermore, as a backup for the computer, the human has to help in detection and diagnosis of system failures. Viewing the human as a monitor, supervisor, and diagnostician leads to three new analogies: ideal observer, time-shared computer, and logical problem solver. The ideal observer analogy, discussed in Chapter 2, is not really that new. However, the recent emphasis on the human's role as a monitor has increased the potential usefulness of this analogy. The usefulness of the time-shared computer analogy, considered in Chapter 4, is related to the desire to represent the human's abilities to supervise multiple tasks. This analogy has continually been embellished as the technology of time-shared computing has developed. The logical problem solver analogy, examined in Chapter 5, is relatively new. Its development reflects a recognition of the importance of understanding the human's role in systems where the ultimate responsibility for system failure is the human's.

The ideal observer, time-shared computer, and logical problem solver analogies cannot yet be viewed as paradigms. Nevertheless, they have received considerable attention and current interests in the human–machine systems area would seem to indicate increased use of these analogies. Two other analogies are considered in this book: information processor and planner, in Chapters 6 and 7, respectively. The current state of human–machine systems research is such that these analogies are not receiving as much attention as the other analogies discussed in this chapter. This is perhaps due to the fact that the analysis tools associated with the information processing and planning analogies are not as well-developed and straightforward as the tools associated with the other analogies. Nevertheless, as discussed in Chapters 6 and 7, these new analogies appear to offer interesting possibilites.

The Notion of Constrained Optimality

We noted in our discussion of analogies that one of the most important reasons for adopting an analogy is the set of analysis tools that may thereby by provided. For example, consider a situation where a servomechanism analogy appears to be appropriate (e.g., a manual control task). By substituting the equations of a servomechanism for the human, one can then use control theory to calculate overall human–machine system performance.

One might further assume that the servomechanism is optimal in the sense that it minimizes mean-squared errors between desired machine output and actual output. Using this approach, one would probably find that the optimal servomechanism performed better (i.e., lower error scores) than possible for the human. From this result, one might conclude that the human is not optimal. However, an alternative perspective is to view the human as optimal within his psychological and/or physiological constraints. Thus, if we design a servomechanism that cannot observe stimuli perfectly, cannot make movements perfectly, and perhaps has less than a perfect internal model (i.e., mental model) of how the machine works and then solve for the optimal performance, we might find that theoretical and empirical results compare more favorably. In fact, such an approach is basic to much of what will be discussed in this book. We shall refer to this idea as *constrained optimality*.

Of course, representing the human as being imperfect in terms of perception, execution, and internal models requires that we quantify these imperfections. Sometimes, deficiencies in perception and execution can be represented as zero-mean random noise processes. Then, one can use the variance of the noise as a free parameter with which to match data.

However, it can be the case that imperfect perception and/or execution is somewhat systematic and cannot be represented as random. For example, it may be possible to misunderstand a display or use the wrong control.

One of the more difficult human constraints to represent is an imperfect internal model. It is certainly more straightforward to assume that the human has a perfect knowledge of how the machine works and then, add various noise sources as constraints. This approach has worked quite well for tasks such as manual control of vehicles which respond relatively fast. However, for slow vehicles, some monitoring tasks, and problem solving tasks, this approach may not be acceptable. Then, one faces the difficult problem of trying to figure out the human's internal representation of the machine with which interaction is required.

At this point, we shall not elaborate further on the use of analogies and the notion of constrained optimality. However, we shall repeatedly return to these concepts in later chapters. These concepts provide one of the central themes of this book.

The Scope of This Book

This book is mainly concerned with the representation and calculation steps of the modeling process. Each chapter will consider a particular methodology in terms of how it can be applied to representing human–machine interaction. These discussions will include presentation of the basic calculation procedures appropriate to each methodology. Thus, the first part of each chapter will include a tutorial on the methodology of interest; the second part will consider applications to human–machine systems.

There are a wide variety of methodologies that qualify as approaches to modeling human–machine interaction. The methodologies considered in this book include estimation theory, control theory, queueing theory, fuzzy set theory, and a few other methods. The theories of estimation, control, and queueing are included because they have a broad, well-developed theoretical base and have been applied to modeling human behavior in a variety of domains. Although control theory is definitely the more ubiquitous, estimation theory is frequently an integral part of many control theory models. Furthermore, with the human becoming more of a monitor or supervisor, rather than in-the-loop controller, the theories of estimation and queueing should be increasingly important approaches to modeling human behavior. While the theory of fuzzy sets has a well-developed theoretical base, the emphasis in this rather new field has been almost solely on theory, and relatively few applications have emerged. However, this situation is definitely changing, and it is quite likely that many fuzzy set models of human behavior will appear within the next several years.

Beyond the broadly based and well-developed methods noted in the last paragraph, this book will also consider a few methods that are somewhat more narrow in their applicability but, nevertheless, quite useful. The rule-based production system models of the artificial intelligence domain, elementary notions from pattern recognition, and also Markov chain models are considered. These particular methods were selected because they have been extensively applied within a variety of interesting, but nevertheless somewhat limited, domains.

Several relatively interesting methods are not discussed in this book. Classical manual control theory has been thoroughly treated by Sheridan and Ferrell (1974) and thus is allocated limited coverage here. Information theory and signal detection theory have also been omitted because they are covered elsewhere [e.g., Sheridan and Ferrell (1974)] and, in addition, because they do not have the robust domains of application that exist for the theories of control, queueing, and so on. Decision theory has not been explicitly included because, with the possible exception of multiattribute utility theory (Keeney and Raiffa 1976), the field is much too broad and nonintegrated to capture succinctly in a single chapter.

Numerous examples of the application of systems engineering models to modeling human–machine interaction are presented in this book. Problem domains considered include aircraft piloting, air traffic control, monitoring and control of industrial processes, detection and diagnosis of system failures, text editing, and several more elementary tasks. Absolutely no attention is devoted to personality, motivation, etc. Hence, this book does not, by any means, address human behavior in the complete sense of the phrase. Nevertheless, the types of behavior considered are among the more important of those that can realistically be studied within engineering design.

Chapter 2
Estimation Theory

Estimation theory is concerned with the problem of dealing with uncertainty. There are basically three types of uncertainty with which humans must cope in human–machine systems: input uncertainty, measurement uncertainty, and parameter uncertainty. Figure 2.1 depicts a human–machine system with input and measurement uncertainty.

Input uncertainty is due to the fact that the human usually does not know exactly how the environment will affect the machine (e.g., wind gusts when driving a car or turbulence when flying an airplane). It is quite typical to view the machine as having two kinds of input. First, of course, the human creates inputs for the machine. The second kind of input comes from the environment and includes deterministic inputs (e.g., measurable head or tail winds) and random inputs such as the wind gusts and turbulence noted above. When inputs are random, the human is in the situation of not knowing exactly what the input will be, and thus will be uncertain about future demands upon the system.

Measurement uncertainty refers to the possibility of the machine's output being corrupted by some random process before the human perceives it as input. This corruption can be due to limitations of the machine such as inaccurate instrumentation and/or can be attributed to uncertainty generated by the human's perceptual process. Either way, the human often must cope with uncertain measurements.

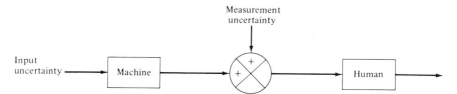

Figure 2.1. Human–machine system with input and measurement uncertainty.

Parameter uncertainty refers to the human's knowledge of the input–output characteristics of the machine (i.e., the human's internal model). It seems quite reasonable to claim that system performance will be enhanced if the human has a good understanding of how the machine works. With this knowledge, the human can anticipate how the machine will respond to an action and, in that way, act more appropriately. Beyond being reasonable, this notion also has a considerable theoretical basis (Conant and Ashby 1970). However, we shall not pursue such issues at this point in our discussion.

The human's knowledge of the machine can be represented as some structural input–output relationship with various free parameters. It is interesting to consider how the human develops this knowledge and further what the implications are if the human has less than perfect knowledge of the machine. These questions will be approached in this chapter by using estimation theory to hypothesize how the human copes with parameter uncertainties.

As a slight digression, let us consider somewhat further the meaning of the phrase "parameter uncertainty" as it refers to the human's understanding of the machine's input–output characteristics. It should be emphasized that the human can, of course, also be uncertain of the structure of the machine's input–output relationship. However, instead of explicitly discussing structural uncertainty, we can view possible internal models as having completely general structures within which most parameters are zero. In that way, an inappropriate structure can be represented by particular parameters being nonzero and thereby, for example, allow inappropriate nonlinear terms into the structure of the internal model. From this perspective, all uncertainty about the input–output characteristics of the machine can be viewed as parameter uncertainties.

Within this chapter, we shall consider the first of many analogies of human behavior. Namely, we shall consider the appropriateness of representing the human as an *ideal observer*. We shall start with the assumption that the human deals with input, measurement, and parameter uncertainties in an optimal manner, subject to various behavioral constraints that we shall later discuss. This assumption will allow us to consider the possibility of the human being an optimal decision maker in estimation tasks.

To be able to discuss estimation theory in sufficient depth, we first must consider some introductory linear systems theory. Then, we shall go on to discuss approaches to dealing with measurement and parameter uncertainty. The tutorial material in this chapter emphasizes linear estimation theory and is based on reviews by Rouse and Gopher (1977) and Rouse (1977a) as well as books by Lee (1954), Meditch (1969), Morrison (1969), and Mendel (1973).

Linear Dynamic Systems

A *system* is an arbitrarily bounded phenomenon (e.g., an airplane, a human, a society, etc.) that has inputs u_1, u_2, . . . , u_r and outputs z_1, z_2, . . . , z_m. In vector form,[1] one can represent the set of inputs by $\mathbf{U} = (u_1, u_2, \ldots, u_r)$ and the set of outputs by $\mathbf{Z} = (z_1, z_2, \ldots, z_m)$.

Knowledge of $\mathbf{U}$ is usually insufficient for determining $\mathbf{Z}$. For example, knowing that a ship changed course by 20 degrees does not provide sufficient information to determine its final heading—one has to know the ship's initial heading. Similarly, knowledge that the ship increased its speed by 5 knots is not sufficient to allow one to determine its final speed—one has to know its initial speed. In fact, continuing our ship example, one would have to know a ship's initial position, heading, and speed to be able to determine where the ship would be if the rudder angle and throttle setting (i.e., the inputs) were changed.

This set of variables that intervene between $\mathbf{U}$ and $\mathbf{Z}$ are termed the *state* variables of the system and are denoted by $\mathbf{X} = (x_1, x_2, \ldots, x_n)$. The value of n is the *order* of the system. For a system of order n, one must have n state variables, knowledge of which along with $\mathbf{U}$ is sufficient to determine $\mathbf{Z}$. However, a variety of choices of state variables are possible. In other words, there is no unique set of variables that must be employed. Usually, as examples in the upcoming pages will illustrate, one tries to choose physically meaningful and easily measurable variables.

To clarify these ideas further, consider the following example. The input of a farm system might be considered to be the amount of fertilizer used, whereas the output of the farm is the total number of bushels of grain produced. Thus this is a single-input ($r = 1$), single-output ($m = 1$) system. Assume that the fertilizer is divided among n plots of ground such that the treatment may vary among plots but not within plots. Then, the state of the farm system is characterized by $\mathbf{X}$, where each state variable x_i represents

---

[1] Boldface letters represent vectors or matrices. Italic letters represent elements of vectors or matrices if subscripted, simple scalars if not.

the number of bushels of grain produced by the ith plot. The output of the system is $z = x_1 + x_2 + \cdots + x_n$. A knowledge of the state is necessary to know how changes in u (fertilizer input) affect z. This point will be discussed further a little later.

A *dynamic* system is one whose state changes with time. Typically, the state at time $t + dt$ is related to the state at time t and the input at time t. Thus, a dynamic system is often considered to have a memory in the sense that its state (or output) at any point in time cannot be predicted solely on the basis of its input. One can relate this idea to human learning in that performance on a particular task depends not only on the parameters of the task but also on the human's previous experience with the task. Thus, learning is a dynamic process.

If a system's state at time $t + dt$ is linearly related to past states and inputs, then the system is termed *linear*. For example, a single-input ($r = 1$), single-output ($m = 1$), first-order linear dynamic system can be represented by[2]

$$x(t + 1) = \phi x(t) + \psi u(t) \tag{2.1}$$

$$z(t + 1) = h x(t + 1), \tag{2.2}$$

where ϕ, ψ, and h are assumed to be constants, although, in general, this need not be true.

Linearity can also be defined using the *superposition principle*. Suppose that an input of $u(t) = u_a(t)$ produces an output of $z(t) = z_a(t)$ while an input of $u(t) = u_b(t)$ produces an output of $z(t) = z_b(t)$. Then, if the system is linear, an input of $u(t) = c_1 u_a(t) + c_2 u_b(t)$ will produce an output of $z(t) = c_1 z_a(t) + c_2 z_b(t)$, where c_1 and c_2 are arbitrary constants. This principle will be of great use later in the discussion.

Considering the farm example, one might utilize (2.1) to represent each of the n plots. Thus, production in year $t + 1$ is related to both production in year t and fertilizer input at the end of year t. Note that even though $u(t)$ may be zero for a particular year (e.g., due to a petroleum shortage), each plot will still produce something due to built-up nutrients in the ground as reflected by last year's production.

One can represent the whole farm by using n versions of equation (2.1) and employing ϕ_i and ψ_i in the equation for the ith plot. Note that $\psi_1 + \psi_2 + \cdots + \psi_n = 1$ reflects the fact that one can only allocate as much fertilizer as is input. Further, note that (2.2) is replaced by $z = x_1 + x_2 + \cdots + x_n$, yielding a single-input, single-output, nth-order linear dynamic system.

---

[2] To simplify notation and avoid having so many symbols, let $dt = 1$. Since the time scale is arbitrary, this simplification does not limit the generality of the results.

To emphasize the difference between system state and output, consider how one might predict future farm outputs $z(t + 1)$, $z(t + 2)$, ... If one had knowledge of the input $u(t + 1)$, $u(t + 2)$, ..., but had no knowledge of the initial state of the system, then the equation representing each plot of land would have two unknowns and could not be solved. In other words, a knowledge of the initial productivity of each plot as well as a knowledge of the fertilizer inputs is needed to predict future production.

Another example that will serve to clarify the difference between system state and output is the longitudinal motion of an automobile traveling on a straight road. In this case, the input is the angle of the accelerator pedal while the output is the position of the vehicle. If the input over some time period $t_o \leqslant t \leqslant t_f$ is known, what is the position of the vehicle at time t_f? Obviously, this question cannot be answered unless the position of the vehicle at t_0 (i.e., its initial position) and the vehicle's velocity at t_0 (i.e., its initial velocity) are known. The state of the system in this example is the vehicle's position and velocity. Thus, system state provides a much more detailed description of the system than available by observing the system output.

Given (2.1) and (2.2), the system's output at time $t + T$ (i.e., T time units into the future) can be determined by considering the system's output at $t + 2, t + 3, \ldots$:

$$x(t + 2) = \phi x(t + 1) + \psi u(t + 1)$$
$$= \phi^2 x(t) + \phi\psi u(t) + \psi u(t + 1),$$
$$x(t + 3) = \phi x(t + 2) + \psi u(t + 2)$$
$$= \phi^3 x(t) + \phi^2 \psi u(t) + \phi\psi u(t + 1) + \psi u(t + 2),$$
$$\vdots$$

and thus,

$$x(t + T) = \phi^T x(t) + \sum_{i=0}^{T-1} \phi^{T-i-1} \psi u(t + i), \tag{2.3}$$

$$z(t + T) = h x(t + T). \tag{2.4}$$

Note that $x(t + T)$ is linearly related to $x(t)$ and $u(t)$, $u(t + 1)$, ... Thus, the fact that nonlinear operations have been performed on the system parameters ϕ and ψ in no way affects the linearity of the system. As a final point on linearity, note that the sequence of system outputs $z(t)$, $z(t + 1)$, ... does not, in general, follow a straight line. Thus, linearity does not imply a straight-line output.

When analyzing a dynamic system, an initial question that one might ask concerns the system's *stability*. An unstable system produces an output that grows continually in time. For linear systems, this phenomenon is unrelated to the system's input. An arbitrarily small perturbation of the system's state may cause it to grow without bound. For example, consider a broomstick that is perfectly balanced in an upright position. Any small perturbation of the stick will cause it to fall. An upright broomstick is an unstable system. Now, consider a broomstick that is attached to some support by a spring. If the broomstick is perturbed, the spring will cause it to return to its upright position. Such a system is stable.

As another example, consider the situation when an animal is introduced to an area where it has no natural predators and food is plentiful. The animal population will continually grow. However, if a predator is added to the system, then the animal population may behave in a manner similar to the spring-supported broomstick and be stable.

A stable system is often defined as one that produces a bounded response to a bounded input. However, even unstable systems produce bounded outputs since the broomstick will hit the floor and the animals may eventually run out of food. Thus, the idea of unbounded response must be accepted as a simplification or abstraction of reality.

In (2.3) and (2.4), as T becomes large, then the absolute value of ϕ must be less than or equal to one (i.e., $|\phi| \leqslant 1$) or $z(t + T)$ will increase in time without bound. In some situations, a value of $\phi = 1$ is also undesirable since the effect of $x(t)$ will never die out in that case.

Equations (2.1) and (2.2) can be generalized to yield for a multi-input, multi-output, nth-order linear dynamic system

$$x_1(t + 1) = \phi_{11} x_1(t) + \cdots + \phi_{1n} x_n(t) + \psi_{11} u_1(t) + \cdots + \psi_{1r} u_r(t),$$

$$x_2(t + 1) = \phi_{21} x_1(t) + \cdots + \phi_{2n} x_n(t) + \psi_{21} u_1(t) + \cdots + \psi_{2r} u_r(t),$$

$$\vdots$$

$$x_n(t + 1) = \phi_{n1} x_1(t) + \cdots + \phi_{nn} x_n(t) + \psi_{n1} u_1(t) + \cdots + \psi_{nr} u_r(t),$$

or, in matrix form,

$$\mathbf{X}(t + 1) = \mathbf{\Phi}\mathbf{X}(t) + \mathbf{\Psi}\mathbf{U}(t), \tag{2.5}$$

and

$$z_1(t + 1) = h_{11} x_1(t + 1) + \cdots + h_{1n} x_n(t + 1),$$

$$z_2(t + 1) = h_{21} x_1(t + 1) + \cdots + h_{2n} x_n(t + 1),$$

$$\vdots$$

$$z_m(t + 1) = h_{m1} x_1(t + 1) + \cdots + h_{mn} x_n(t + 1),$$

or, again in matrix form,

$$\mathbf{Z}(t + 1) = \mathbf{HX}(t + 1). \tag{2.6}$$

Equations (2.5) and (2.6) could be used to represent the single-input, single-output farm example. In that case, $r = 1$ and $m = 1$ while $h_{11} = h_{12} = \cdots = h_{1n} = 1$. Equations (2.1) and (2.2) can also be represented by (2.5) and (2.6) by letting $r = 1$, $m = 1$, and $n = 1$. Thus (2.5) and (2.6) represent the general case.

As an example of a multi-input, multi-output system, one can consider an automobile where the two inputs are steering wheel angle and accelerator pedal angle while the outputs are lateral and longitudinal position. Also, the farm example might be expanded to include both fertilizer and rainfall as inputs and multiple crops as outputs (e.g., corn and soybeans) and, in that way, yield a multi-input, multi-output system.

The response of the system described by (2.5) and (2.6) is similar in form to (2.3) and (2.4) and given by

$$\mathbf{X}(t + T) = \Phi^T \mathbf{X}(t) + \sum_{i=0}^{T-1} \Phi^{T-i-1} \Psi \mathbf{U}(t + i), \tag{2.7}$$

$$\mathbf{Z}(t + T) = \mathbf{HX}(t + T). \tag{2.8}$$

The stability of the system described by (2.7) and (2.8) depends on Φ as it depended on ϕ in the first-order case. The criterion for stability with respect to Φ is analogous to the criterion with respect to ϕ (i.e., $|\phi| \leqslant 1$) but mathematically it is more complex and beyond the scope of this book (Takahashi et al. 1970, pp. 132–136).

The type of system response which has been of concern thus far is termed the *transient* response. One can see that, as T becomes large in (2.3) and (2.7), a stable system will yield a response that depends only on the input and is insensitive to the initial state of the system. The response for large T is called the *steady-state* response. If the input $\mathbf{U}(t)$ does not change with time, then the steady-state response will not change with time. On the other hand, if $\mathbf{U}(t)$ changes continually, then so will the system output. However, for large T, ϕ^T will become very small and, thus, these changes will be independent of the initial state of the system.

Recalling the notion that a dynamic system is one that remembers its initial conditions and past inputs, steady-state response can be defined as that point at which the system has forgotten its initial conditions. Thus, a dynamic system has a fading memory or, one might say, a memory window.

To consider steady-state response to a time-varying $\mathbf{U}(t)$, first consider how a scalar time function $u(t)$ can be represented in general. Assume that

$u(t)$ is periodic with period T. That is, $u(t)$ repeats itself every T units of time. Alternatively, one can assume interest in $u(t)$ only over the interval $0 \leqslant t \leqslant T$, in which case whether or not $u(t)$ repeats itself is not important.

Knowing T, one might try to approximate $u(t)$ by $u(t) \cong a_0 + a_1 \cos(\omega t) + b_1 \sin(\omega t)$ where $\omega = 2\pi/T$. In general, this approximation is not satisfactory. Thus, the approximation requires more terms. If more cosine and sine terms are to be added, the frequency (i.e., ω) cannot be less than $2\pi/T$ since $u(t)$ is assumed to repeat itself every T. Thus, the frequency must exceed $2\pi/T$. Since any new terms in the approximation should be uncorrelated with the terms already in use, new terms that are orthogonal to the current terms should be chosen. This requires that cosine and sine terms whose frequencies are harmonics of the fundamental frequency $\omega = 2\pi/T$ are chosen. Therefore, the frequencies $2\omega, 3\omega, \ldots$ are used, and the approximation becomes

$$u(t) = a_0 + \sum_{k=1}^{\infty} [a_k \cos(k\omega t) + b_k \sin(k\omega t)]. \qquad (2.9)$$

Thus, the approximation includes frequencies $\omega, 2\omega, 3\omega, \ldots$ and amplitudes $a_1, b_1, a_2, b_2, a_3, b_3, \ldots$. It is important to note that, for any physically realizable $u(t)$, a_k and b_k will approach zero for large k. In other words, physically realistic systems cannot produce signals of infinite frequency.

Equation (2.9) is very useful since, due to the previously noted superposition principle, the output of a linear system whose total input can be represented by a sum of inputs is the sum of the outputs due to each individual component of the input. Thus, knowing the response of a linear system to $u(t) = b \sin(\omega t)$ for all values of ω is sufficient to characterize the response of the system to any periodic input.

If this discussion were to continue in the context of discrete systems as described by (2.5) and (2.6), the input to be considered would be sampled values of $b \sin(\omega t)$. One might then compute the response and discuss stability criteria. Unfortunately, this would be a very awkward discussion since the frequency response of a system [i.e., response to $b \sin(\omega t)$ as a function of ω] is usually only considered for continuous systems—those systems described by differential equations.

Instead of resorting to discussion of continuous-time systems, it will simply be noted that the output of a linear system, whose input is $u(t) = b \sin(\omega t)$, is $z(t) = g(\omega)b \sin[\omega t + \alpha(\omega)]$ where g is the *gain* and α is the *phase angle*, both of which depend on ω. The gain $g(\omega)$ is the ratio of output amplitude to input amplitude. This is often expressed in terms of decibels (dB) which is $20 \log_{10}[g(\omega)]$. Thus, for example, if output amplitude is the same as input amplitude, the gain is 0 dB.

The phase angle $\alpha(\omega)$ is related to a translation in time. Thus, $b \sin(\omega t)$ and $b \sin(\omega t + \alpha)$ differ in that they peak and cross the zero axis at different times while the amplitudes and frequencies (or periods) are the same. A good example of the concept of phase angle is that due to the human's reaction time delay. When a human reacts to an unpredictable stimulus, the response is delayed. In tasks where the human's response can be suitably modeled as a periodic function, this reaction time delay will appear as a phase lag. Specifically, $\alpha(\omega)$ will equal $-\omega\tau$ where τ is the reaction time delay.

A continuous-time linear system is often represented by $g(\omega)$ and $\alpha(\omega)$ functions or plots of $g(\omega)$ in dB and $\alpha(\omega)$ in degrees versus ω on semilogarithmic graph paper. Such a graph is called a Bode plot. This set of input–output amplitude ratios and phase shifts represents the *transfer function* of the system. Transfer functions are often used to represent single-input, single-output, constant coefficient linear systems. For multi-input, multi-output linear systems, the *state model* of (2.5) and (2.6) is more appropriate, especially if the coefficients of the model are time-varying.

One can also characterize the response of a nonlinear system to sinusoidal inputs. For certain types of nonlinearity, a sufficient input–output description is provided by the ratio of amplitude of the output to the amplitude of the input. This approximation is called a *describing function*. Within the human–machine systems domain, that portion of the input–output relationship not accounted for by the describing function is often attributed to a random process and is termed *remnant*.

To consider stability in the frequency domain, one must differentiate between *open-loop* and *closed-loop* stability. An open-loop system is one whose inputs are not affected by its outputs. On the other hand, a closed-loop system is one whose inputs are affected by its outputs. For example, an automobile without a driver is an open-looped system whereas, once a driver is added, the loop becomes closed because the driver's steering wheel angle is related to the automobile's deviation from the desired path.

Consider a system that is open-loop stable. Since it is stable, its response will reach steady state and its frequency response can be characterized with $g(\omega)$ and $\alpha(\omega)$. Now, suppose that the system's output is "fed back" so that it becomes a component of the system's input. Will the system be closed-loop stable?

It is difficult to discuss the answer to this question rigorously without going way beyond the scope of this book. Thus, the stability criterion will simply be presented. This will at least enable the reader to understand the jargon by which stability is discussed.

The stability criterion in the frequency domain is that $g < 1$ when $\alpha = -180°$. The amount that g is less than one (i.e., $1 - g$) when $\alpha = -180°$ is termed the *gain margin*, while the amount that α is greater than $-180°$ when $g = 1$ is termed the *phase margin*.

Before moving on to a new topic, we shall summarize our discussion of linear dynamic systems. First the concept of a system and its inputs, outputs, and state were introduced. Linear dynamic systems were defined as those whose future states are linearly related to their past inputs and initial conditions. System response and stability for single-input, single-output and multi-input, multi-output systems were considered. Finally, steady-state frequency response was discussed.

Comparing the methods discussed here with more traditional behavioral science methodologies, one can see that the focus of the engineering model is the system and its state space. This model is constructed to account for how the system's past inputs and initial conditions affect its present and future outputs. If the system's past inputs and initial condition are not relevant, then direct (or static) input–output equations will suffice to describe the system's behavior fully. Such direct input–output equations are familiar to the behavioral scientist. For example, such equations form the basis of analysis of variance.

Thus, time is the dominant dimension in the system dynamics model. Notions of causality and periodicity are also important aspects of this model. Periodicity especially manifests itself in the concept of steady-state response (although the steady-state response need not be periodic). Steady state response might represent a desired level or goal of the system, or perhaps a dynamic balance that depends on the nature of the inputs and the characteristics of the system. Steady-state response also represents the ability of a stable system to, in effect, forget its past. Thus, a dynamic system is conceptualized as having a fading or sliding-window memory whose width depends on the characteristics of the specific dynamic system.

The behavioral scientist or engineer is most likely to benefit (directly or indirectly) from the quantitative techniques discussed in this chapter when perceiving research problems in terms of phenomena such as those described in the above two paragraphs.

State Estimation

Now, expanding the system description of (2.1) and (2.2),

$$x(t + 1) = \phi x(t) + \psi u(t) + \gamma w(t), \qquad (2.10)$$

$$z(t + 1) = h x(t + 1) + v(t + 1), \qquad (2.11)$$

where w is a zero-mean Gaussian disturbance[3] with variance σ_w^2, and v is a zero-mean Gaussian measurement noise with variance σ_v^2. Although w and

---

[3] The disturbance input is sometimes used to reflect uncertainty in ϕ, ψ, etc., and not just an "input" in the usual sense.

v are assumed to be completely random processes, methods are available for considering processes that violate these assumptions (Meditch 1969, pp. 192–199).

Equations (2.10) and (2.11) allow the representation of a much wider class of phenomena than is possible using (2.1) and (2.2). Equations (2.10) and (2.11) represent deterministic plus random inputs as well as imperfect measurements of the state of the system.

Given a set of measurements $z(t + 1)$, $z(t + 2)$, ..., $z(t + i)$, estimation involves generating an estimate of $x(t + j)$, denoted by $\hat{x}(t + j | t + i)$, in order to minimize some function of the estimation error denoted by the criterion $L = L[\tilde{x}(t + j)]$, where $\tilde{x}(t + j) = x(t + j) - \hat{x}(t + j | t + i)$.

There are three types of estimation problem. When $j > i$, the estimation problem is called *prediction*. When $j = i$, the problem is called *filtering*. When $j < i$, the problem is termed *smoothing*. Thus, prediction is the task of estimating what will happen in the future. On the other hand, filtering and smoothing occur after the data have been observed to remove the uncertainty due to measurement noise. Smoothing is similar to a moving weighted average, whereas filtering is like the task of smoothing the end points of a sample.

To determine the optimal estimate, assume that the criterion L is scalar, assigns no penalty for zero estimation error, does not decrease for an increased estimation error, and is symmetric about zero. The estimate that minimizes the expected value of L is given by

$$\hat{x}(j|i) = E[x(j)|z(1), z(2), \ldots, z(i)], \tag{2.12}$$

where $E[\cdot]$ denotes the expected value.[4] Thus, for a wide class of criteria, (2.12) holds and one need not be too concerned with the particular functional form of L.

Before the specific results for prediction, filtering, and smoothing are discussed, we shall consider ignoring u, the deterministic input. Since u is assumed known or specified, it does not affect the uncertainty of the state of the system. Thus, u can be considered in a deterministic manner which will, by the way, also allow consideration of the mean of w if it happens to be nonzero. The procedure for considering deterministic inputs within an estimation context is discussed by Meditch (1969, 132–135). This discussion will proceed assuming that $u = 0$.

Another point which should be mentioned concerns why x is being predicted instead of z. First, one can always obtain $z = hx$, and thus, both estimates are available. More importantly, when we consider systems whose

---

[4] The use of the expected value reflects the statistical nature of the problem. In fact, the underlying theory is in many ways similar to linear regression, particularly for the Gaussian disturbance and measurement noise assumed within the discussions here. One conceptual difference is that linear regression is directly based on minimizing expected error squared. However, as clearly illustrated by Lee (1964), the approach taken in this chapter and linear regression result in the same estimation algorithms.

order n exceeds the number of outputs m, the estimate of the state gives more information than the estimate of the output. This additional information is essential if future system outputs are to be predicted. Furthermore, when control theory is considered in Chapter 3, the reader will find that this information is of considerable value.

For prediction one employs (2.12) with (2.3) and (2.10) and obtains[5]

$$\hat{x}(j|i) = \phi^{j-i} x(i). \tag{2.13}$$

However, it is unlikely that $x(i)$ will actually be available. Instead, an estimate of the state at time i is employed; this yields

$$\hat{x}(j|i) = \phi^{j-i} \hat{x}(i|i). \tag{2.14}$$

Thus, the prediction is based on a filtered estimate which leads to consideration of filtering.

Assume that $j = i + 1$ and that $\hat{x}(i|i)$ is available. Before $z(i + 1)$ is observed, the best estimate of $x(i + 1)$ can be determined using (2.14); it yields

$$\hat{x}(i + 1|i) = \phi \hat{x}(i|i). \tag{2.15}$$

When $z(i + 1)$ is observed, one has two estimates of $x(i + 1)$, namely the prediction and the observation. One must decide how much each of these estimates is to be trusted. Referring to (2.10) and (2.11), note that σ_v^2 is crucial to the decision of how to trade off the two estimates. If σ_v^2 is small, then one has confidence in $z(i + 1)$. On the other hand, if σ_v^2 is large, one has more confidence in $\hat{x}(i + 1|i)$.

The results of trading off the two estimates will be the filtered estimate $\hat{x}(i + 1|i + 1)$. A linear combination of the two estimates can be formed to yield

$$\hat{x}(i + 1|i + 1) = \hat{x}(i + 1|i) + k(i + 1)[z(i + 1) - h\hat{x}(i + 1|i)], \tag{2.16}$$

where k is a filter coefficient or gain that allows one to trade off the prediction and observation.

One would intuitively like k to behave as follows. If σ_v^2 is small, then k should approach $1/h$ so that the filtered estimate approaches the observation. If σ_v^2 is large, then k should approach zero so that the filtered estimate approaches the prediction. [Notice how substituting $k = 0$ or $k = 1/h$ into (2.16) produces the desired effect.]

Whether or not σ_v^2 is large depends on its contribution to the uncertainty in $z(i + 1)$ relative to the uncertainty inherent in the prediction $\hat{x}(i + 1|i)$. Using (2.10),

$$\sigma_{\tilde{x}}^2(i + 1|i) = \phi^2 \sigma_{\tilde{x}}^2(i|i) + \gamma^2 \sigma_w^2, \tag{2.17}$$

where $\sigma_{\tilde{x}}^2(i + 1|i)$ is the variance of the prediction error and $\sigma_{\tilde{x}}^2(i|i)$ is the

---

[5] Recall that w is a zero-mean process, and thus $E[w(t)] = 0$ for all values of t.

variance of the filtering error. Defining $\tilde{z}(i + 1|i) = z(i + 1) - \hat{z}(i + 1|i)$ as the error in predicting $z(i + 1)$, then

$$\sigma_{\tilde{z}}^2(i + 1|i) = h^2 \sigma_{\tilde{x}}^2(i + 1|i) + \sigma_v^2. \tag{2.18}$$

Now, using (2.18), one can form a filter gain with the desirable properties and obtain

$$k(i + 1) = \frac{1}{h} \cdot \frac{h^2 \sigma_{\tilde{x}}^2(i + 1|i)}{[h^2 \sigma_{\tilde{x}}^2(i + 1|i) + \sigma_v^2]}. \tag{2.19}$$

Thus, the weighting on each of the two estimates is inversely proportional to the portion of error variance attributable to each estimate. If $\sigma_v^2 \gg h^2 \sigma_{\tilde{x}}^2(i + 1|i)$, then one relies on the prediction. On the other hand, if $\sigma_v^2 \ll h^2 \sigma_{\tilde{x}}^2(i + 1|i)$, then one relies on the observation. This result should be intuitively acceptable to the reader.

The filter will be completely defined once it is determined how to calculate $\sigma_{\tilde{x}}^2(i|i)$. Using (2.16), one finds that

$$\sigma_{\tilde{x}}^2(i|i) = [1 - k(i)h]^2 \sigma_{\tilde{x}}^2(i|i - 1) + k^2(i)\sigma_v^2, \tag{2.20}$$

which, with additional manipulation, leads to

$$\sigma_{\tilde{x}}^2(i|i) = [1 - k(i)h]\sigma_{\tilde{x}}^2(i|i - 1). \tag{2.21}$$

Thus, (2.16), (2.17), (2.19), and (2.21) define the optimal filter which is commonly known as the Kalman filter. The same equations can be derived from a least-squares fit approach and, for the more general case about to be considered, a multiple linear regression approach will yield an identical estimation scheme (Lee 1964, pp. 49–56).

Equations (2.5) and (2.6) for multi-input, multi-output nth-order linear dynamic systems can be extended in a manner analogous to the extension of (2.1) and (2.2) to obtain

$$\mathbf{X}(t + 1) = \mathbf{\Phi}\mathbf{X}(t) + \mathbf{\Psi}\mathbf{U}(t) + \mathbf{\Gamma}\mathbf{W}(t) \tag{2.22}$$

$$\mathbf{Z}(t + 1) = \mathbf{H}\mathbf{X}(t + 1) + \mathbf{V}(t + 1), \tag{2.23}$$

where $\mathbf{W}$ is a vector with elements $w_1, w_2, \ldots, w_p$ and $\mathbf{V}$ is a vector with elements $v_1, v_2, \ldots, v_m$. The covariance of $\mathbf{W}$ is denoted by $\mathbf{Q}$ while the covariance of $\mathbf{V}$ is denoted by $\mathbf{R}$.

The optimal prediction for this general case is

$$\hat{\mathbf{X}}(j|i) = \mathbf{\Phi}^{j-1}\hat{\mathbf{X}}(i|i) \tag{2.24}$$

and the optimal filter is defined by

$$\hat{\mathbf{X}}(i + 1|i + 1) = \hat{\mathbf{X}}(i + 1|i) + \mathbf{K}(i + 1)[\mathbf{Z}(i + 1) - \mathbf{H}\hat{\mathbf{X}}(i + 1|i)], \tag{2.25}$$

where

$$K(i + 1) = P(i + 1|i)H'[HP(i + 1|i)H' + R]^{-1}, \tag{2.26}$$

$$P(i + 1|i) = \Phi P(i|i)\Phi' + \Gamma Q \Gamma', \tag{2.27}$$

$$P(i + 1|i + 1) = [I - K(i + 1)H]P(i + 1|i), \tag{2.28}$$

where I denotes an identity matrix and $P(i + 1|i)$ and $P(i|i)$ are the multidimensional equivalents of $\sigma_{\tilde{x}}^2(i + 1|i)$ and $\sigma_{\tilde{x}}^2(i|i)$, respectively. P is usually initialized as $P(0|0) = E[X(0)X(0)']$.

Now let us consider the smoothing problem of producing $\hat{x}(j|i)$ for $j < i$. Since smoothing is not central to much of our later discussions, we shall only briefly review one algorithm. We shall consider fixed-point smoothing [as opposed to fixed-interval or fixed-lag smoothing; see Meditch (1969, Chap. 5)]. The optimal smoothed estimates for the general case of (2.22) and (2.23) are given by

$$\hat{X}(j|i) = \hat{X}(j|i - 1) + B(i)[\hat{X}(i|i) - \hat{X}(i|i - 1)], \tag{2.29}$$

where

$$B(i) = \prod_{k=j}^{i-1} P(k|k)\Phi'P(k + 1|k)^{-1}, \tag{2.30}$$

$$P(j|i) = P(j|i - 1) - B(i)[P(i|i) - P(i|i - 1)]B(i)', \tag{2.31}$$

and P is the estimation error covariance with initial condition $P(j|j)$ obtained from the filtered estimates.

There are two things that are important to note about this smoothing algorithm. First, (2.29) is very similar in structure to (2.25). Second, the optimal smoother is employed after the optimal filter has been utilized which, as noted earlier, requires the use of the optimal predictor. Thus, the general state estimation problem involves prediction, filtering, and smoothing. Although we shall not pursue smoothing algorithms further here, the interested reader can find fairly thorough treatments in Lee (1964) and especially Meditch (1969).

To employ the estimation algorithms outlined in this chapter, one needs a knowledge of Φ, Γ, H, Q, and R, as well as $P(0|0)$. It is interesting to consider the impact of errors in one's knowledge of these parameters. The effect of errors in Q, R, and $P(0|0)$ are discussed by Heffes (1966). The possibility of estimator instability is discussed by Fitzgerald (1971) as it relates to inaccurate knowledge of systems parameters. Athans (1971) considers increasing Q to account for inaccuracies in Φ, and Morrison (1969) discusses this effect in more detail.

Mehra suggests the use of correlation procedures for identification of **Q** and **R** (1970) and also for estimating Φ, Γ, **H**, and **R** simultaneously (1971). We should note here that Γ and **Q** cannot be uniquely identified without initial knowledge of one of them. However, this does not present difficulties, since Γ and **Q** can enter the estimation algorithms without being separated.

The point of this brief discussion on assumptions is to point out that violation of assumptions does not render the algorithms inoperable and that there are methods for adaptively correcting errors.

It is interesting to compare the procedures for prediction, filtering, and smoothing presented here with the way such procedures are defined and employed in most behavioral science research. With time being the independent dimension, the estimation methods presented in this chapter can be viewed as measurement procedures for dealing with different points along the time continuum. In the static (nondynamic) statistical methods employed in most behavioral research, the varied dimension is some scale of input values or magnitude levels. As the history of system output is not considered, direct input–output (stimulus–response) functions are computed. Regression techniques are used for prediction, whereas averaging techniques are employed to filter or smooth data corrupted by measurement noise. Time-dependent constructs are most commonly avoided. A notable exception to this generalization is research in learning theory (Krantz et al. 1974) where time dependencies are, of course, essential phenomena. In fact, the Markov models employed by many learning theorists are based on the same field of mathematics from which a considerable portion of estimation theory as discussed here was developed (Meditch 1969, Chap. 4).

Parameter Estimation

Earlier, we briefly considered how one's knowledge of Φ, Γ, etc. affects the state estimation algorithms. If one is to use these algorithms to describe human behavior in estimation tasks, then one should consider how the human gains knowledge of these parameters. In this section, we shall consider algorithms designed to "learn" Φ, Γ, and so on. This topic is termed parameter estimation or system identification. Our development will follow that of Lee (1964) until we reach topics beyond the scope of his text.

Lee considers the case of identifying a system that can be modeled as

$$\mathbf{X}(t + 1) = \Phi \mathbf{X}(t) + \Psi \mathbf{U}(t) + \Gamma \mathbf{W}(t), \tag{2.32}$$

$$\mathbf{Z}(t + 1) = \mathbf{H} \mathbf{X}(t + 1). \tag{2.33}$$

Equations (2.32) and (2.33) can be transformed first to canonical form and then to difference equation form to yield, for single-input, single-output systems,

$$z(t) = \sum_{i=1}^{n} -a_i z(t - i) + b_i u(t - i) + c_i w(t - i), \tag{2.34}$$

where the system defined by (2.32) and (2.33) must be observable. Equation (2.34) can be expressed more compactly as[6]

$$z(t) = \mathbf{S}(t - 1)'\phi + \mathbf{T}(t - 1)'\mathbf{C}, \tag{2.35}$$

where

$$\mathbf{S}(t - 1)' = [z(t - n) \cdots z(t - 1)u(t - n) \cdots u(t - 1)]$$

and

$$\mathbf{T}(t - 1)' = [w(t - n) \cdots w(t - 1)],$$

whereas

$$\phi' = [-a_n \cdots - a_1 b_n \cdots b_1] \text{ and } \mathbf{C}' = [c_n \cdots c_1].$$

The estimate $\hat{\phi}$ that minimizes the expected value of the squared estimation error resulting with $\hat{\phi}$ is given by

$$\hat{\phi}(t + 1) = \hat{\phi}(t) + \mathbf{D}(t)[z(t - 1) - \mathbf{S}(t)'\hat{\phi}(t)], \tag{2.36}$$

where

$$\mathbf{D}(t) = \mathbf{P}(t)\mathbf{S}(t)[\mathbf{S}(t)'\mathbf{P}(t)\mathbf{S}(t) + 1]^{-1}, \tag{2.37}$$

$$\mathbf{P}(t) = \mathbf{P}(t - 1) - \mathbf{D}(t - 1)\mathbf{S}(t - 1)'\mathbf{P}(t - 1), \tag{2.38}$$

and $\mathbf{P}(0)$ is often assumed arbitrarily large with $\hat{\phi}(0) = 0$.

The above algorithm requires that no more than one element of $\mathbf{C}$ is nonzero. If $\mathbf{C}$ has more than a single nonzero element, then the residuals of the estimation process will be correlated, and the algorithm will yield biased estimates. This difficulty can be avoided by only updating $\hat{\phi}$ after every n inputs, because $\mathbf{C}$ can have at most n nonzero elements. Thus, (2.36)–(2.38) become

$$\hat{\phi}[(t + 1)n] = \hat{\phi}(tn) + \mathbf{D}(tn)[z(tn + 1) - \mathbf{S}(tn)'\hat{\phi}(tn)], \tag{2.39}$$

where

$$\mathbf{D}(tn) = \mathbf{P}(tn)\mathbf{S}(tn)[\mathbf{S}(tn)'\mathbf{P}(tn)\mathbf{S}(tn) + 1]^{-1}, \tag{2.40}$$

$$\mathbf{P}(tn) = \mathbf{P}[t(n - 1)] - \mathbf{D}[t(n - 1)]\mathbf{S}[t(n - 1)]'\mathbf{P}[t(n - 1)]. \tag{2.41}$$

---

[6] The use of ϕ as a matrix does not agree with our usual upper case notation for matrices. The lower case has been adopted here to avoid confusion with Φ while still emphasizing the relationship between ϕ and Φ.

If there is no control input [i.e., $u(t) = 0$ for all t], then the identification algorithms specified in (2.36)–(2.38) or (2.39)–(2.41) are still appropriate with $\mathbf{S}$ and ϕ redefined to exclude the control input and associated parameters.

Now consider the situation when (2.33) is more appropriately expressed as

$$\mathbf{Z}(t + 1) = \mathbf{H}\mathbf{X}(t + 1) + \mathbf{V}(t + 1). \tag{2.42}$$

Astrom and Eykoff (1971) discuss how biased estimates of ϕ (and hence Φ) result when such measurement noise is present. They suggest ways of avoiding this problem. For example, assuming the system order n to be arbitrarily larger than the actual system order helps to reduce biases.

Landau's model reference adaptive approach (1974) is structurally similar to (2.36)–(2.38) and avoids the bias due to measurement noise by using $\hat{\mathbf{Z}}(t - 1)$ to obtain $\hat{\mathbf{Z}}(t)$. The limitation here is that the model reference method assumes a knowledge of the input.

Now, we want to consider estimation algorithms that put more stress on recent inputs than on older inputs. In other words, we want to discuss estimators that forget.

Following Morrison (1969), suppose that our estimation algorithm for filtering is presented with a vector of observations $\mathbf{Z}(1), \mathbf{Z}(2), \ldots, \mathbf{Z}(j)$ where we assume (2.32) and (2.42) as a system model with the additional assumption, for the moment, that $\mathbf{Q} = 0$. If our estimate of the current state of the process is $\hat{\mathbf{X}}(j)$, then we can form an error vector

$$\mathbf{E}(j) = \begin{bmatrix} \mathbf{Z}(j) & -\mathbf{H}\hat{\mathbf{X}}(j|j) \\ \mathbf{Z}(j - 1) & -\mathbf{H}\Phi(j - 1,j)\hat{\mathbf{X}}(j|j) \\ \cdot \\ \cdot \\ \cdot \\ \mathbf{Z}(1) & -\mathbf{H}\Phi(1,j)\hat{\mathbf{X}}(j|j) \end{bmatrix} \tag{2.43}$$

where $\Phi(i,j)$ represents a transition from time i to time j, in this case a transition backward in time. If we weight $\mathbf{E}(j)$ by $\tilde{\mathbf{W}}(j)' = [\mathbf{W}(j)\mathbf{W}(j - 1) \cdots \mathbf{W}(1)]$,[7] then our weighted error vector is $\tilde{\mathbf{W}}(j)'\mathbf{E}(j)$. We would like to find the estimate $\hat{\mathbf{X}}(j|j)$ that minimizes $[\tilde{\mathbf{W}}(j)'\mathbf{E}(j)]'[\tilde{\mathbf{W}}(j)'\mathbf{E}(j)]$. Further, we would like to choose $\tilde{\mathbf{W}}(j)$ such that this jth order vector can be a simple function of the $(j - 1)$st order $\tilde{\mathbf{W}}(j - 1)$ given by

$$\tilde{\mathbf{W}}(j) = \begin{bmatrix} - & \dfrac{1}{\beta\tilde{\mathbf{W}}(j - 1)} & - \end{bmatrix}, \tag{2.44}$$

---

[7] This is somewhat different from Morrison's approach in that he weights $\mathbf{E}(j)'\mathbf{E}(j)$.

where $0 < \beta \leqslant 1$. In this way, when $\beta < 1$, the weightings on old errors will asymptotically approach zero and the estimator will forget old errors.

With such a weighting function, Lee (1964) and Mendel (1973) derive an identifier that forgets. Weighting *state* estimation errors, in the fashion prescribed by (2.44), results in (2.36)–(2.38) becoming

$$\hat{\phi}(t + 1) = \hat{\phi}(t) + \mathbf{D}(t)[z(t + 1) - \mathbf{S}(t)'\hat{\phi}(t)], \qquad (2.45)$$

where

$$\mathbf{D}(t) = \mathbf{P}(t)\mathbf{S}(t)[\mathbf{S}(t)'\mathbf{P}(t)\mathbf{S}(t) + \beta^2]^{-1}, \qquad (2.46)$$

$$\mathbf{P}(t) = \beta^{-2}[\mathbf{P}(t - 1) - \mathbf{D}(t - 1)\mathbf{S}(t - 1)'\mathbf{P}(t - 1)]. \qquad (2.47)$$

Thus we have separately discussed state estimators and parameter estimators, the latter with fading memory. In reality, the human is faced with simultaneous estimation of states and parameters. Algorithms for such situations have appeared in the literature (Bar-Shalom 1972; Nelson and Stear 1976). Bar-Shalom (1972) utilizes a smoother and then an identifier, whereas Nelson and Stear (1976) employ an identifier and then a filter. They both solve simultaneous estimation as two separate problems. We shall illustrate a similar approach in a later discussion.

Applications

The state and parameter estimation algorithms presented in this chapter are useful for representing the human decision maker as an *ideal observer* in the task of monitoring dynamic processes. Appropriate monitoring tasks might involve explicit prediction, filtering, or smoothing of the states of the process. Alternatively, state estimation might only be implicit in an overall task such as detection of process failures. In this latter case, in order to apply the algorithms considered here, one must infer the existence of some type of estimation mechanism within the human.

A further condition necessary for applying the estimation algorithms presented in this chapter is that the dynamic process of interest (defined by Φ, Γ, etc.) be known or at least identifiable. This condition is usually easily satisfied for tasks such as driving an automobile, flying an airplane, or piloting a ship. However, for tasks such as managing an insurance company, one can easily see that difficulties arise. In fact, we need not even resort to such extreme examples to illustrate this point. In many relatively well-defined situations such as monitoring in power plants or in fluid process control plants, it is difficult to derive the equations of the process in a form readily amenable for use by the algorithms presented here. In

Chapter 6, we shall consider how one might model human decision making in situations where the process dynamics are unknown or ill-defined. However, in this chapter, we limit our discussions to situations satisfying the known or identifiable dynamics condition.

A Failure Detection Model

Consider a task where the human monitors a dynamic process described by

$$X(t + 1) = \Phi X(t) + \Gamma W(t), \tag{2.48}$$

$$Z(t + 1) = HX(t) + V(t), \tag{2.49}$$

where W and V are zero-mean Gaussian processes with covariances Q and R, respectively. The human's task is to detect when (2.49) becomes

$$Z(t + 1) = HX(t) + M(t) + V(t), \tag{2.50}$$

where M is an arbitrary deterministic time function which represents a process "failure."

Gai and Curry (1976a) have developed a model of the human decision maker in this task. A diagram of the model appears in Figure 2.2. The model assumes that the human is an ideal observer who employs a Kalman filter to eliminate uncertainty due to V, and then employs a detection mechanism based on sequential analysis. Notice that V is attributed to

Figure 2.2. Gai and Curry's Failure Detection Model.
Based on Gai and Curry (1976a).

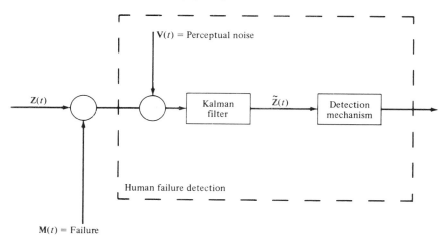

uncertainty generated in the perceptual process. In a manner similar to the
Weber–Fechner law of classical psychophysics, it is assumed that the
variance of the perceptual noise is proportional to the variance of the
displayed signal.

The failure detection model assumes that the human's decisions are
based on the *residuals* of the Kalman filter. The residual at time t is given by

$$\tilde{\mathbf{Z}}(t) = \mathbf{Z}(t) - \hat{\mathbf{Z}}(t|t-1)$$
$$= \mathbf{Z}(t) - \mathbf{H}\hat{\mathbf{X}}(t|t-1) \qquad (2.51)$$
$$= \mathbf{Z}(t) - \mathbf{H}\Phi\hat{\mathbf{X}}(t-1|t-1).$$

Thus, the residual is the difference between one's expectations [i.e.,
$\mathbf{H}\Phi\hat{\mathbf{X}}(t-1|t-1)$] and what actually occurred [i.e., $\mathbf{Z}(t)$]. This comparison
can be likened to that used for the derivation of the Kalman filter where
predictions and observations were combined to produce an appropriate
filter.

If no failure has occurred [i.e., $\mathbf{M}(t) = 0$], then $\mathbf{Z}(t)$ can be described as
a sequence of independent observations of a zero-mean Gaussian process.
However, if a failure occurs [i.e., $\mathbf{M}(t) \neq 0$], then $\mathbf{Z}(t)$ will no longer be a
zero-mean process. Gai and Curry postulated a mechanism whereby the
human detects this change of mean.

We shall consider the case of scalar observations where the failed mode
can be characterized by a step change in mean of θ. The decision function
at time T is given by

$$\tilde{\lambda}(T) = \sum_{t=1}^{T} [\tilde{z}(t) - \theta/2]. \qquad (2.52)$$

Assuming $\theta > 0$, $\tilde{\lambda}(T)$ will tend to be negative when no failure has
occurred and positive when a failure has occurred. Since the human need
not report the absence of errors, the decision function can be modified to
yield

$$\lambda(T) = \begin{cases} \tilde{\lambda}(T) & \text{if } \tilde{\lambda}(T) > 0 \\ 0 & \text{if } \tilde{\lambda}(T) \leqslant 0. \end{cases} \qquad (2.53)$$

Using this decision function, Gai and Curry show that a failure should
be reported if

$$\lambda(T) > LN(A)/\theta, \qquad (2.54)$$

where

$$A = (1 - P_{MS})/P_{FA},$$
$$P_{MS} = \text{probability of miss},$$
$$P_{FA} = \text{probability of false alarm}.$$

This model was compared to the performance of two human subjects whose task was detection of failures characterized by step and ramp changes in $m(t)$. It was found that the mean time to detection (i.e., the average value of T when a failure was reported) of the model compared favorably with that of the subjects. Further, the model was also successfully applied to describing human behavior in monitoring the progress of an automatic landing system for an aircraft (Gai and Curry 1976b). Finally, the model has been extended to the problem of detecting process changes (i.e., changes in Φ) (Curry and Govindaraj 1977). Thus, this model has been shown to have some degree of general applicability.

It is interesting to consider how this model differs from what one would develop if an automated detection algorithm were desired. Except for the addition of perceptual noise, the model presented here is an optimal decision maker. Thus, one can conclude that the human decision maker, for the class of failure detection tasks considered here, can be adequately described as an optimal decision maker subject to the constraint that his perceptions of stimuli are corrupted by noise. As indicated in Chapter 1, this notion of the human as the optimal subject to various psychological and/or physiological constraints will appear repeatedly throughout this book.

A General Model for Estimation Tasks

In many situations, the human is required to estimate the state of a dynamic process. For example, to drive a car successfully, purchase stocks and bonds, or plan a research project, the human must predict future states of various dynamic processes. Similarly, in order to avoid having to react to short-term random patterns superimposed on long-term trends, the human's observations must be smoothed to obtain an estimate of the state of the underlying dynamic process.

To aid the human in such tasks, several investigators have developed *predictor displays* that provide the human with a forecast of the future states of the dynamic process of interest (Sheridan and Ferrell 1974, pp. 268–273). Such displays usually improve the performance of the human–machine system. This raises the question of why humans are less than optimal predictors.

Based on a series of studies (Rouse 1972,1973,1976,1977a), Rouse has proposed that inadequate knowledge of the dynamics of the process (i.e., Φ, Γ, etc.) may be the cause of this suboptimality. This hypothesis led to the development of a theory of how humans learn about Φ or, in other words, develop an internal model of the dynamics of a process.

The proposed theory exploits several notions of mathematical psychology. It is assumed that humans develop both short- and long-term memory models of the dynamic process of interest. Both memories are assumed to decay exponentially at rates β_s and β_l for short- and long-term memories, respectively. The possibility of perceptual noise is also considered.

More formally, it is assumed that the human is monitoring dynamic processes described by

$$\mathbf{X}(t + 1) = \mathbf{F}[\mathbf{X}(t), \mathbf{U}(t), \mathbf{W}(t), t + 1], \tag{2.55}$$

$$\mathbf{Z}(t + 1) = \mathbf{H}[\mathbf{X}(t + 1), \mathbf{V}(t + 1), t + 1], \tag{2.56}$$

where $\mathbf{X}$, $\mathbf{U}$, $\mathbf{W}$, and $\mathbf{V}$ are as defined earlier while $\mathbf{F}$ and $\mathbf{H}$ are arbitrary vector-valued functions. The argument $t + 1$ is included to allow for the possibility of $\mathbf{F}$ and $\mathbf{H}$ being time-varying relationships.

Regardless of whether the human's estimation task is prediction, filtering, or smoothing, the following scheme was proposed for the adaptation of *each* of the short- and long-term memory models. The first phase is prediction, which is accomplished by using

$$\hat{\mathbf{X}}(t + i|t) = \hat{\Phi}^i \hat{\mathbf{X}}(t|t), \qquad i = 1, 2, \ldots, T. \tag{2.57}$$

Upon observing $\mathbf{Z}(t + i)$, $i = 1, 2, \ldots, T$, the human then filters using

$$\hat{\mathbf{X}}(t + i|t + i) = \hat{\mathbf{X}}(t + i|t + i - 1) + \mathbf{K}(t + i)$$
$$\times [\mathbf{Z}(t + 1) - \mathbf{H}\hat{\mathbf{X}}(t + i|t + i - 1)] \tag{2.58}$$

for $i = 1, 2, \ldots, T$ and then may smooth (especially if the task is smoothing) using

$$\hat{\mathbf{X}}(t + i|t + j) = \hat{\mathbf{X}}(t + i|t + j - 1) + \mathbf{B}(t + j)$$
$$\times [\hat{\mathbf{X}}(t + j|t + j) - \hat{\mathbf{X}}(t + j|t + j - 1)] \tag{2.59}$$

for $i = 1, 2, \ldots, T$ and $j = i + 1, i + 2, \ldots, T$ and finally identifies

$$\hat{\phi}(t + i) = \hat{\phi}(t + i - 1) + \mathbf{D}(t + i - 1)$$
$$\times [\mathbf{H}\hat{\mathbf{X}}(t + i|t + i) - \hat{\mathbf{S}}(t + i - 1)'\hat{\phi}(t + i - 1)] \tag{2.60}$$

for $i = 1, 2, \ldots, T$. The gain matrices $\mathbf{K}$, $\mathbf{B}$, and $\mathbf{D}$ are as defined earlier and reflect appropriate uses of the memory parameters β_s and β_l.

Assuming a single short-term model and a single long-term model, then, for any particular task, the human internally produces $\hat{\mathbf{X}}_s(t + i|t)$ and $\hat{\mathbf{X}}_l(t + i|t)$ and then externally produces

$$\hat{\mathbf{X}}(t + i|t) = \alpha \hat{\mathbf{X}}_l(t + i|t) + (1 - \alpha)\hat{\mathbf{X}}_s(t + i|t), \tag{2.61}$$

where α is a parameter that reflects the human's trade-off between short- and long-term memory. Thus, the theory employs three parameters: β_s, β_l, and α. The memory parameters reflect constraints on the human's ability, whereas the trade-off parameter reflects (perhaps conscious) decision making on the part of the human. The theory is summarized in Figure 2.3.

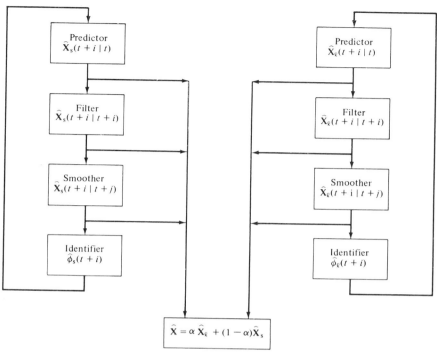

Figure 2.3. Rouse's model for estimation tasks.
Based on Rouse (1977a).

Rouse compared this theory to experimental data in two ways. First, he considered whether or not the parameters of the theory could be manipulated so that they matched the estimation error scores achieved by subjects in three different experiments. All of the experiments involved displaying to the subject the outputs of a dynamic process describable by (2.22) and (2.23), namely an nth-order, linear dynamic process. In the first experiment, four subjects were asked to predict the time series of future states of the process (Rouse 1972). This is termed a multistage prediction task. In the second experiment, eight subjects performed a single-stage prediction task in that they were asked to forecast one time unit into the future (Rouse 1973). Finally, a third experiment entailed having four subjects perform a multistage smoothing task (Rouse 1976). Since the models developed for these three tasks were all special cases of the general theory reviewed here, Rouse concluded that the theory would do quite well in terms of comparisons of theoretical and empirical estimation errors.

Rouse's second approach to comparing theory and data involved contrasting the time histories of behavior produced by the theory and experimental subjects. He considered eight different dynamic processes which varied considerably in terms of their predictability (i.e., the portion of the output variance that was predictable based on a knowledge of the dynamics). Data from four subjects in the previously mentioned single-stage prediction task were used. For each dynamic process, the theory was employed to produce approximately 100 single-stage predictions. The memory parameters were fixed to be $\beta_s = 0.9$ and $\beta_l = 1.0$ whereas α was varied, independently for each subject, in an effort to match the time histories of predictions of the theory and each subject. It was found that trials with predictable dynamics resulted in relatively low values of α, whereas trials with less predictable dynamics resulted in larger values of α. This is a quite reasonable result because it means that subjects based their predictions on recent trends for predictable processes but used long-term statistics for less predictable processes. Thus, the theory allows a behaviorally meaningful interpretation of how short- and long-term models affect decision making in estimation tasks.

Considering the comparison of time histories in more detail, the correlation between the sequence of predictions of the theory and each subject ranged from approximately 0.98 for predictable processes to approximately 0.60 for very unpredictable processes. For processes where at least 10% of the variance of their outputs were attributable to the dynamics, rather than to the random inputs, the correlations were typically above 0.90. Thus, the theory is reasonably good in terms of describing human behavior in prediction tasks where the process of interest is at least somewhat predictable.

Summary

In this chapter, we have considered the problem of modeling how humans cope with uncertain dynamic environments. An ideal observer analogy was adopted and optimal state as well as parameter estimation algorithms were derived. In the course of presenting these algorithms, we defined and explained the concepts of system, state, linearity, response, and stability.

Noting that applying estimation theory as presented in this chapter requires that the dynamic process of interest be known or identifiable, we presented two applications which served as our first detailed illustrations of the use of analogies and the notion of constrained optimality. The ideal observer analogy was found to work quite nicely for describing human behavior in a failure detection situation and in a variety of explicit estimation tasks. Within the failure detection scenario, the human was assumed to be an optimal decision maker subject to the constraint of having a noisy perceptual process. For the estimation tasks, the human was assumed optimal subject to memory limitations and the potential of inappropriate trade-offs between estimates based on short- and long-term memory. These two different formulations of constrained optimality serve to point out the possibility that constraints and thus behavioral interpretations may not be unique. This difficulty will be discussed in Chapter 7.

Chapter 3
Control Theory

In Chapter 2, we were concerned with the human's abilities to observe a dynamic process and then predict its future states or perhaps detect that the dynamic and/or statistical properties of the process had changed. Figure 2.1 was used to depict this human–machine system.

In this chapter, we are going to complicate the human's task somewhat. The human must observe the process and estimate its states, but also must generate a control action in an attempt to cause the states of the process to assume desirable values. This human–machine control situation is depicted in Figure 3.1.

Notice that Figure 3.1 was created by rearranging Figure 2.1 and then making two additions. First, we "closed the loop" by making the human's output affect the machine. Second, we provided the human with an error measure in terms of desired output minus actual output. This second addition is necessary in order to provide the human with a criterion by which to control the system, namely, minimize errors or some function of errors.

For the task depicted in Figure 3.1, we can view the human as an error-nulling device. In control theory terms, we can say that the human acts as a *servomechanism*. Using this servomechanism analogy, we can proceed to analyze the human–machine system as if the human were in fact a

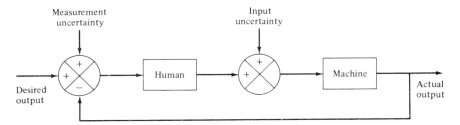

Figure 3.1. Human–machine control system.

servomechanism. As noted earlier, employing analogies such as this in order to expedite analysis is a crucially important aspect of the engineering approach to analysis of human–machine systems.

Within this chapter, we shall first review the basic elements of control theory. Most of the tutorial material is based on a review by Rouse and Gopher (1977) as well as books by Takahashi et al. (1970) and Meditch (1969).

Closing the Loop

Returning to (2.1) and (2.2), consider system control, i.e., choosing $u(t)$. The purpose of controlling a system is to force its response [(2.3) and (2.4)] to behave in some desirable manner. Thus, for example, in an automobile one chooses steering wheel angle in order to force the automobile to stay in the appropriate lane. Also, one chooses accelerator pedal angle in order to maintain the desired speed. Clearly, the particular choices of steering wheel angle and accelerator pedal angle are related to both the characteristics of the highway and the current state of the vehicle $x(t)$. If $u(t)$ is a function of $x(t)$, then one has *feedback* or *closed-loop* control. [Alternatively, $u(t)$ may be a function of the system's output $z(t)$ if the human does not have access to the state $x(t)$. Later in this chapter, we shall consider how the human might reconstruct $x(t)$ from observation of $z(t)$.]

In the automobile driving example (and other situations), one is not totally free to choose $u(t)$ because the highway also provides some input. This difficulty can usually be avoided by defining the state of the vehicle relative to the highway. When such a transformation is appropriate, it does not limit the generality of the results. However, we shall later consider situations where one would not want to employ this transformation.

Since the human only directly observes the output of the system $z(t)$, rather than its state $x(t)$, we shall first consider a control of the form

$$u(t) = cz(t), \tag{3.1}$$

where c is some constant. Substituting (3.1) into (2.1) and (2.2) yields

$$x(t + 1) = (\phi + \psi ch)x(t). \qquad (3.2)$$

If one specifies that the system should respond as would a system with parameter ϕ_d, then

$$\phi_d = \phi + \psi ch \qquad (3.3)$$

and

$$c = (\phi_d - \phi)/\psi h. \qquad (3.4)$$

Consider the effect of the gain c. Since $z(t)$ represents deviations from a desired trajectory (e.g., the highway), one would want to choose c in order to decrease $x(t + 1)$ and hence $z(t + 1)$. Since ϕ, ψ, and h are usually positive, the goal of decreasing $z(t + 1)$ requires that c be negative. Thus, from (3.4) we see that $\phi_d < \phi$ and, as the magnitude of c increases, $\phi_d - \phi$ becomes larger in magnitude. Therefore, increasing the magnitude of the gain c is equivalent to making the system respond as if its parameter ϕ_d was smaller and smaller. As we noted in Chapter 2, as ϕ decreases, ϕ^T decreases more quickly and thus the system more quickly "forgets" its past and responds to new inputs. From this perspective, increasing the magnitude of the gain is equivalent to increasing the responsiveness of the system. However, there is a limit to how much the magnitude of c can be increased because we must also consider the stability criterion $|\phi| \leqslant 1$. Applying this criterion to ϕ_d, (3.3) gives us the limit on c. From this brief discussion, one should be able to see how choosing c relates to trading off the responsiveness and stability characteristics of a system. This is a basic trade-off with which a human controller must deal.

Applying the notions upon which (3.4) is based to the control of multi-input, multi-output, nth-order systems defined by (2.5) and (2.6), one obtains

$$\mathbf{C} = \mathbf{\Psi}^{-1}[\mathbf{\Phi}_d - \mathbf{\Phi}]\mathbf{H}^{-1}, \qquad (3.5)$$

which requires that $\mathbf{\Psi}^{-1}$ and $\mathbf{H}^{-1}$ exist. This implies that n inputs and n outputs must exist. For situations where this assumption cannot be satisfied, see Takahashi (1970, Chap. 10).

Optimal Control

Instead of specifying desired system characteristics (via ϕ_d or Φ_d), one can attempt to find the control that minimizes some criterion. A typical choice is minimization of the quadratic criterion given by

$$J = E\left[\sum_{i=1}^{T} az^2(t + i) + bu^2(t + i - 1)\right], \qquad (3.6)$$

where $E[\cdot]$ denotes the expected value and is included to reflect the possible statistical nature of the problem. This criterion expresses a desire to minimize squared deviations of system output from the desired output [i.e., $z(t) = 0$] while also minimizing squared control effort necessary to achieve this goal. The ratio a/b allows one to trade off deviations of system output and control effort.[1]

Relating (3.6) to the farm example discussed earlier, assume that one is interested in finding a government subsidy policy [the control $u(t)$] that achieves some desired total farm production in the United States [the output $z(t)$]. The desired total farm production is such that deviations from that desired value results in scarcity and high prices (due to underproduction) or abundance and low prices (due to overproduction). Let $z(t)$ be the deviations from the desired farm production. The choice of a and b in (3.6) is dictated by how willing one is to subsidize farming to achieve the desired production. If a/b is large, then one can expect that the optimal control will employ large $u(t)$ in an effort to minimize $z^2(t)$. On the other hand, if a/b is small, then the result will be that the optimal control does not employ large $u(t)$ because minimization of government expenditures is more important (e.g., a balanced budget) than achieving the desired farm production.

Considering again the automobile driving example, the criterion specified by (3.6) is intuitively quite appealing. If a/b is large, then the optimal control would be to use many control actions in order to keep the automobile *exactly* on the desired path at the desired speed. However, if a/b is small, then the optimal control would avoid or minimize control activity and accept the resulting deviations from the desired path and speed. In this case, perhaps the work load associated with frequent control actions would be beyond what one would be willing to withstand in order to stay exactly on the desired path.

Before considering the solution to the optimal control problem based on the criterion function of (3.6), it is useful to note how the use of a cost function differs from specifying ϕ_d or Φ_d. Basically the criterion function allows one to specify performance goals rather than system characteristics. It would seem that performance goals are a more direct means for specifying objectives since the choice of system characteristics requires the analyst to transform ϕ_d or Φ_d into performance characteristics, which

---

[1] Notice that $z(t)$ and $u(t + T)$ do not appear in (3.6). This reflects the fact that the optimal control is being computed at time t and thus, $z(t)$ cannot be affected by $u(t)$, $u(t + 1)$, etc. Further, since the criterion only considers the interval from t to $t + T$, $u(t + T)$ is not of concern because it only affects $z(t + T + 1)$, $z(t + T + 2)$, etc. Although various alternative formulations of the criterion in (3.6) are available (e.g., with limits from $i = 0$ to $i = T - 1$), they all reflect the same ideas and produce the same results.

necessitates considerable insight and experience. However, while specifying performance goals is more direct, the choices of a and b are not usually easy. One must still resolve the fundamental trade-off between effort and results. Thus, one should look at optimal control formulations as useful transformations of a problem rather than elimination of a problem.

The solution to the optimal control problem results in

$$u(t + i) = c(t + i)x(t + i), \tag{3.7}$$

and thus one has a linear control law. The way in which c is found can be related to a constrained multiple regression problem where the constraint is that u and x are also related by (2.1). Such a dynamic constraint requires a somewhat sophisticated optimization method and therefore only the solution will be presented here.

The gain c is given by

$$c(t + i) = -\psi d(t + i + 1)\phi/[\psi^2 d(t + i + 1) + b], \tag{3.8}$$

where

$$d(t + i) = \phi^2 d(t + i + 1) + \phi d(t + i + 1)\psi c(t + i) + h^2 a \tag{3.9}$$

and $d(t + T) = 0$. If T is large, approaching infinity, then d becomes constant and thus

$$u(t + i) = cx(t + i), \tag{3.10}$$

$$c = -\psi d\phi/(\psi^2 d + b), \tag{3.11}$$

where d is the solution of

$$d^2\psi^2 + d(b - \phi^2 b - \psi^2 ah^2) - abh^2 = 0, \tag{3.12}$$

such that $c < 0$.

Although it may not be obvious to the reader, (3.11) and (3.12) are such that large a/b yields a large c (i.e., large magnitude, but still $c < 0$) whereas small a/b yields a small c. Within the context of the above farm example, the gain c shows how government expenditures in terms of farm subsidies should react to deviations in farm output from the desired output. For the automobile example, the gain c shows how steering wheel angle and accelerator pedal angle should react to deviations from the desired path and speed. This behavior is what was hypothesized in the discussion of the impact of a and b on the resulting optimal control.

For a multi-input, multi-output nth-order system, (3.6) becomes

$$J = E\left[\sum_{i=1}^{T} \mathbf{Z}(i)'\mathbf{A}\mathbf{Z}(i) + \mathbf{U}(i - 1)'\mathbf{B}\mathbf{U}(i - 1)\right], \tag{3.13}$$

where $t = 0$ for notational convenience. The optimal control is given by

$$\mathbf{U}(i) = \mathbf{C}(i)\mathbf{X}(i), \tag{3.14}$$

where

$$\mathbf{C}(i) = -[\boldsymbol{\Psi}'\mathbf{D}(i+1)\boldsymbol{\Psi} + \mathbf{B}]^{-1}\boldsymbol{\Psi}'\mathbf{D}(i+1)\boldsymbol{\Phi} \tag{3.15}$$

and

$$\mathbf{D}(i) = \boldsymbol{\Phi}'\mathbf{D}(i+1)\boldsymbol{\Phi} + \boldsymbol{\Phi}'\mathbf{D}(i+1)\boldsymbol{\Psi}\mathbf{C}(i) + \mathbf{H}'\mathbf{A}\mathbf{H}, \tag{3.16}$$

with $\mathbf{D}(T) = 0$. As T approaches infinity, $\mathbf{D}$ approaches a steady-state value so that

$$\mathbf{U}(i) = \mathbf{C}\mathbf{X}(i) \tag{3.17}$$

$$\mathbf{C} = -(\boldsymbol{\Psi}'\mathbf{D}\boldsymbol{\Psi} + \mathbf{B})^{-1}\boldsymbol{\Psi}'\mathbf{D}\boldsymbol{\Phi}, \tag{3.18}$$

where $\mathbf{D}$ is the solution of

$$\mathbf{D} = \boldsymbol{\Phi}'\mathbf{D}\boldsymbol{\Phi} - \boldsymbol{\Phi}'\mathbf{D}\boldsymbol{\Psi}(\boldsymbol{\Psi}'\mathbf{D}\boldsymbol{\Psi} + \mathbf{B})^{-1}\boldsymbol{\Psi}'\mathbf{D}\boldsymbol{\Phi} + \mathbf{H}'\mathbf{A}\mathbf{H}. \tag{3.19}$$

The above optimal control solutions are not as intuitively clear as the solutions for the estimation problems. However, two points are especially important to realize. First, the optimal control is a linear function of the state variables. Second, to control optimally, one must employ all of the state variables. As an example, if a human is to control the position of a vehicle, not only the vehicle's present position but also its present velocity must be known. If, for some reason, one is prevented from sensing velocity, one will be unable to control the vehicle optimally.

If the system of interest is described by (2.22) and (2.23), then one has a combined estimation and control problem. Fortunately, there exists a *separation principle* which has been shown, under certain conditions, to allow one to solve the estimation and control problems separately (Lee 1964, pp. 131–133; Meditch 1969, pp. 345–362; Takahashi et al. 1970, pp. 675–690). Thus, one can construct a Kalman filter that generates $\hat{\mathbf{X}}$ from measurements of $\mathbf{Z}$ and then one can employ an optimal controller that utilizes $\hat{\mathbf{X}}$. Note that the filter equations (2.25)–(2.28) are independent of $\mathbf{A}$ and $\mathbf{B}$ in (3.13), whereas the optimal control equations (3.13)–(3.16) are independent of $\mathbf{Q}$ and $\mathbf{R}$ utilized in the filter. Also note that the filter enables full state feedback even when all of the states cannot be measured directly.

Applications

Within this section, we shall discuss several models of human behavior in
control tasks. This topic is called *manual control* (as opposed to automatic
control) and is covered in great detail in other books (Johannsen et al. 1977;
Sheridan and Ferrell 1974; Kelley 1968). The goal of our discussions within
this section is to review several models that have played key roles in the
study of manual control and also to consider various new modeling notions
that have recently been developed.

The control algorithms presented in this chapter are useful for representing the human as a *servomechanism* in the task of controlling continuous
dynamic processes where the most important performance criterion is
minimization of deviations of the state of the process from a desired
trajectory. As with the estimation algorithms of Chapter 2, applying these
control algorithms requires that the dynamics of the process be known or
identifiable.

We shall assume that the human's task is to control a process described
by

$$X(t + 1) = \Phi X(t) + \Psi U(t) + \Gamma W(t), \qquad (3.20)$$

$$Z(t + 1) = HX(t + 1) + V(t + 1). \qquad (3.21)$$

Thus, the human must observe Z (e.g., deviations from desired heading,
deviations from desired velocity)[2] and produce U (e.g., steering wheel angle,
joystick deflection). For over 30 years, various investigators have been
developing models that relate human input (i.e., Z) to output (i.e., U). In
general, these models have taken the same form as (3.20) and (3.21), except
that the input to the human is Z and the output is U. Thus, the human has
been modeled as a linear dynamic system.

Many of these manual control models have *not* been formulated as
discrete processes such as represented by (3.20) and (3.21). Instead, several
models have been expressed in terms of transfer functions, such as
discussed in Chapter 2, with the gain and phase angle at various frequencies
characterizing the input–output relationships of interest (i.e., machine and/
or human). Any of these models that are discussed here have been
converted from their original forms to equivalent forms consistent with the
notation in this book. Therefore, many readers may find that the formulations in the original papers and texts bear little resemblance to those

---

[2] Note that this assumes X to represent deviations from the desired state $X = 0$. Thus, the
coordinate system is not absolute, but instead is referenced to a desired trajectory in an
absolute coordinate system. As mentioned earlier, this transformation usually does not affect
the generality of our results. Exceptions to this rule will be noted later.

presented here. However, it should be stressed that these formulations are conceptually equivalent and allow one to avoid differential equations, convolution integrals, Laplace transforms, and so on.

Identifying the Human Controller

Before considering several manual control models, we shall briefly consider a more direct approach to describing human control behavior. If one were to perform a manual control experiment in which the input signals seen by subjects [i.e., $z(1)$, $z(2)$, . . .] and the control signals produced by subjects [i.e., $u(1)$, $u(2)$, . . .] were recorded, then one could attempt to identify the relationship between u and z. In other words, instead of developing a model and then comparing its behavior to that of the human, one could start by looking at human behavior and identifying appropriate relationships between controls and inputs. While this approach sounds very straightforward, it does, of course, require that one have appropriate empirical data. Such data may be unavailable if the system in which one is interested has not yet been built. This is a typical situation in engineering design.

Nevertheless, direct identification can be quite useful. For example, identification can be used to validate models after data have been collected. In this case, the transfer function (i.e., input–output relationship) of the model can be compared to the transfer function identified from the empirical data. As another example, identification is quite helpful in behavioral studies. For instance, one might like to know how alcohol or lack of sleep affect human behavior in manual control tasks. Identification of transfer functions from data collected for situations with and without alcohol, or with and without adequate sleep, might allow one to determine how the alcohol or lack of sleep affected the human's input–output relationship.

The problem of identifying the input–output characteristics of the human controller has received considerable attention. Sheridan and Ferrell (1974, Chaps. 10 and 11) review some of the more classical, as opposed to modern, methods. Many of the newer methods are discussed by Sage and Melsa (1971) and Eykhoff (1974). Even books that are not particularly devoted to identification often discuss the topic because of important relationships among identification, estimation, and control [e.g., Anderson and Moore (1979)].

Eykhoff's lengthy book is very thorough in dealing with a wide variety of identification problems, many of which involve significantly more complicated situations than are addressed in this chapter. For example, he treats both time-varying and nonlinear situations which require more elaborate algorithms than are presented here. Eykhoff also briefly discusses applications to human–machine systems.

The body of literature on identification is quite extensive and highly applicable to analysis of human–machine interaction. However, we shall not attempt to review all of this material here. Instead, we shall briefly discuss two fairly simple methods of identification. First, we shall consider how the parameter estimation algorithms presented in Chapter 2 can be used to provide a very simple, yet powerful, method of identification. Then, we shall discuss a reasonably new method for dealing with identification, especially for nonlinear systems.

We shall consider the problem of identifying the input–output characteristics of the human plus controlled process. (In other words, we shall *not* attempt to identify the input–output relationship of the human and/or machine independent of one another.) We shall be concerned with situations where the human plus controlled process can be represented by

$$\mathbf{X}(t + 1) = \Phi \mathbf{X}(t) + \Gamma \mathbf{W}(t), \tag{3.22}$$

$$\mathbf{Z}(t + 1) = \mathbf{H}\mathbf{X}(t + 1). \tag{3.23}$$

Note that $\mathbf{U}(t)$ does not appear because it is assumed that the human generates a control of the form $\mathbf{U}(t) = \mathbf{C}\mathbf{X}(t)$ which is coalesced into Φ.

Enstrom and Rouse (Enstrom 1976; Enstrom and Rouse 1977) addressed the problem of identifying Φ, Γ, and so on in a real-time on-line manner by using the fading memory parameter estimation algorithm summarized by (2.45)–(2.47). For manual control with a variety of dynamic processes, they found that a second-order model (i.e., $n = 2$) was adequate and its parameters could be rapidly and accurately estimated using a memory coefficient of $\beta = 0.90$. Using a DEC System-10 and a FORTRAN program, they were able to analyze 5 minute's worth of data in approximately 10 cpu seconds. Thus, the method is readily amenable to implementation on any minicomputer that is no worse than 30 times slower than a DEC System-10. For slower minicomputers, assembly language programming might still allow successful implementation of this identification algorithm.

Enstrom and Rouse employed this method to study the impact of secondary tasks on the human's input–output characteristics. Using randomly occurring mental arithmetic problems as a secondary task, they found that the fading memory estimation algorithm combined with an elementary pattern recognition algorithm (see Chapter 6) was quite successful in detecting when the human shifted attention between controlling and solving arithmetic problems.

One can envision this relatively simple identification method as being of use in two ways in particular. First, as noted earlier, it could be used in a variety of behavioral investigations of the effects of various side tasks, stimulants, depressants, stresses, and so on. A second potential use is in the area of human–computer interaction. Using Enstrom and Rouse's method, a computer could sense in real time whether or not the human was distracted from the control task. If it detected such a distraction, the computer might warn the human against, for example, falling asleep at the steering wheel. Alternatively, the computer might infer that an aircraft pilot is involved in an emergency and therefore, automatically provide assistance in the control task.

Inooka and Inoue (1978) also considered the problem of identifying the input–output characteristics of the human plus controlled process. They employed the group method of data handling (GMDH) approach of Ivankhenko (1971). This method works in the following way. We start by assuming that the human's control $u(t)$ is related to past system states $x(t-1), x(t-2), \ldots, x(t-n)$ in some manner denoted by

$$u(t) = f[x(t-1), x(t-2), \ldots, x(t-n)], \tag{3.24}$$

where $u(t-1)$, $u(t-2)$, and so on do not appear in the right-hand side of (3.24) because we assume that they are also related to previous system states.

To identify $f[\cdot]$, we start by partitioning the empirical data into two sets, one for training (i.e., estimating $f[\cdot]$) and one for testing the fit of the resulting $f[\cdot]$. Selecting pairs of independent variables, $x(t-i)$ and $x(t-j)$, one uses the training data to identify (via linear regression) candidate functions

$$u_k(t) = a_0 + a_1 x(t-i) + a_2 x(t-j) + a_3 x(t-i)^2$$

$$+ a_4 x(t-j)^2 + a_5 x(t-i)x(t-j) \tag{3.25}$$

for $i = 1, 2, \ldots, n$, $j = 1, 2, \ldots, n$, and $k = 1, 2, \ldots, \frac{1}{2}n(n+1)$. Each candidate function is then evaluated in terms of mean-squared error using the testing data. A threshold is used to eliminate candidate functions. For example, one might retain the n candidates with the lowest mean-squared errors. For illustrative purposes, assume that $u_k(t)$, $k = 1, 2, \ldots, n$, happen to be the functions that are retained. Using these functions, one forms another set of candidate functions

$$\tilde{u}_k(t) = b_0 + b_1 u_i(t) + b_2 u_j(t) + b_3 u_i(t)^2$$

$$+ b_4 u_j(t)^2 + b_5 u_i(t)u_j(t), \tag{3.26}$$

where $i = 1, 2, \ldots, n$, $j = 1, 2, \ldots, n$, and $k = 1, 2, \ldots, \frac{1}{2}n(n+1)$.

This iterative process can continue, with nonlinear models of higher and higher order resulting. One stopping rule is to terminate iteration once any candidate function reaches a desired level of mean-squared error. This function, and naturally all its ancestors, is called the GMDH model.

Inooka and Inoue (1978) compared the nonlinear GMDH approach to linear approaches, such as those discussed in Chapter 2, using data from two manual control experiments involving control of a second-order linear dynamic process. Nonlinearities were introduced by having the amplitude of the control device limited in order to produce a saturation effect. The two experiments differed in terms of the level of saturation or, in other words, the extent of the nonlinearity. The GMDH algorithm worked at least as well as the linear algorithm for all cases and, significantly better than the linear algorithm for the highly nonlinear case.

Now that we have briefly considered two identification methods, we shall move on to the problem of a priori modeling. In other words, we shall consider models that are first developed on the basis of reasonable behavioral assumptions and appropriate mathematical theory and then, compared to empirical data. For such models, the structure of the model is determined before data collection and, in fact, often dictates the nature of the experiments performed. The only statistical "fitting" involved with these models involves at most, free parameters within the predetermied structure.

An Intuitive Model

To develop an intuitive model, we shall consider the case of a single input $z(t)$ and a single output $u(t)$. [Recall that $z(t)$ is the input to the human while $u(t)$ is the output produced by the human.] As an initial model, we might simply adopt (3.1). Thus, our initial model would be

$$u(t) = cz(t). \tag{3.27}$$

However, there are several reasons why this model is inadequate. First of all, the human requires a finite amount of time to react to stimuli. For tasks such as controlling a single-output process, this reaction time τ is typically 0.15–0.20 seconds. Thus, the human's output at time t will be related to the observation at time $t - \tau$. Equation (3.27) therefore becomes

$$u(t) = cz(t - \tau). \tag{3.28}$$

Although this modified model is somewhat more realistic, further modifications are necessary. The human's neuromotor system prohibits the instantaneous movement of limbs. For example, if $z(t - \tau)$ suddenly increases dramatically in magnitude, the human will not be able to produce a correspondingly sudden increase in $u(t)$. This limitation can be included by changing (3.28) as follows:

$$u(t) = a_1 u(t - dt) + a_2 z(t - \tau). \tag{3.29}$$

Thus, the human's present control output is related to the control output an instant (dt) earlier and the observation a reaction time earlier. Typical values of a_1 are 0.15–0.20, whereas a_2 is mainly related to the gain c.

Fortunately, the human is able to partially compensate for the limitations of reaction time and neuromotor sluggishness. One way in which this compensation can be achieved is by increasing the gain c. Another way is by anticipating $z(t)$. This can be accomplished by observing the rate of change of $z(t - \tau)$ and extrapolating this rate of change to estimate future values of $z(t)$. This type of behavior can be added to (3.29) to yield

$$u(t) = a_1 u(t - dt) + a_2 z(t - \tau) + a_3 z(t - \tau - dt), \qquad (3.30)$$

where a_1 and a_2 are defined as above and a_3 is related to the gain c as well as to the way in which the human extrapolates $z(t - \tau)$.

Thus, our intuitive model views the human as having two inherent limitations (reflected in τ and a_1) for which the human compensates by increasing the gain (reflected in a_2 as well as a_3) and also by anticipating changes in $z(t - \tau)$. In the 1950s, McRuer and Krendel developed such a model (McRuer and Krendel 1957; Sheridan and Ferrell 1974, pp. 223–230). Given the parameters describing the controlled process (i.e., Φ, Ψ, Γ, etc.) this model can be used to predict the output of the overall human–machine system. Various performance measures such as mean-squared values of $z(t)$ can also be predicted.

The difficulty with this model is that its parameters (i.e., a_1, a_2, and so on) vary considerably with Φ, Ψ, Γ, and so on. The model's parameters are also affected by the characteristics of the input $w(t)$. Thus, a large tabulation of model parameters versus Φ, etc. is necessary, and whenever a new situation is encountered (i.e., a different process for the human to control), new experiments are necessary to determine the model parameters. A more desirable manual control model would be one whose parameters are not quite so task sensitive.

The Crossover Model

McRuer and his colleagues solved this problem in the 1960s by developing the crossover model (McRuer et al. 1965; Sheridan and Ferrell 1974, pp. 231–238). This model includes both the human and the controlled process. The idea motivating this model was that the human adapts the overall human–machine system to have certain "good servo" or, in other words, good stability and response characteristics. Thus, the parameters of the overall human–machine model are much less sensitive to the task than they were with the earlier intuitive model. On the other hand, because we are assuming that the human changes personal characteristics in order to produce desirable system characteristics, the parameters of the human's input–output relationship would, of course, still be quite task sensitive. However, as long as we limit our perspective to the overall human–machine system, we do not have to concern ourselves with these intervening parameters.

A discrete equivalent of the crossover model is

$$z(t) = z(t - dt) + cz(t - \tau_e), \qquad (3.31)$$

where τ_e is an equivalent delay related to both human reaction time and neuromotor response. The parameters τ_e and c are approximately linearly related to single parameter characterizations of the controlled process and frequency characteristics of the input. Note that (3.31) is *not* a model of the human alone, but instead is a model of the human plus controlled process which reflects the adaptation of the human's own dynamics in order to cause the overall dynamics of the human–machine system to have desirable characteristics. Thus, (3.31) is similar to (3.2) in that, once the control $u(t)$ has been defined, it can be substituted into the system equations and thereby eliminated.

With this model, tabulations of parameters are no longer necessary because (3.31) and the linear equations for the parameters τ_e and c completely specify the model over a reasonable range of conditions. Unfortunately, the crossover model is difficult to apply to multi-input, multi-output control tasks. Further, the crossover model is cumbersome when the dynamics of the controlled process are time-varying.

The Optimal Control Model

Based on the developments in estimation theory and optimal control theory that were discussed in Chapter 2 and this chapter, Kleinman, Baron, and Levison developed a new manual control model in the late 1960s and early 1970s (Kleinman et al. 1971; Sheridan and Ferrell 1974, Chap. 12). The basic idea of the model is that the well-motivated, highly trained human is an optimal controller subject to personal limitations. Thus, referring to (3.20) and (3.21), the human must estimate the system state $\mathbf{X}$ using noisy observation $\mathbf{Z}$, where the noise characterized by $\mathbf{V}_z$ is attributed to the classical Weber–Fechner law. Given this estimate, the human must then predict τ into the future to counteract the reaction time delay. Finally, the human generates control outputs $\mathbf{U}$ in order to minimize a criterion similar to (3.13), with the exception that the rate of change of the control (rather than the control itself) is penalized. This change introduces the sluggishness of the human's neuromotor system. Further, the model also allows for motor noise, with covariance $\mathbf{V}_u$ proportional to the variance of the control output, in the sense that the human is psychophysically limited from producing $\mathbf{U}$ perfectly. The optimal control model is summarized in Figure 3.2.

The discrete time equations for this model can be drawn directly from the earlier discussion. The estimation portion of the task is performed using a Kalman filter,

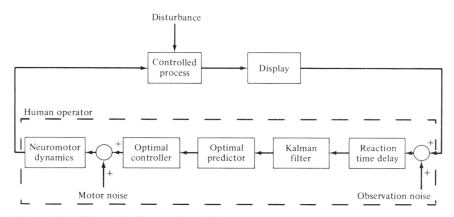

Figure 3.2. The optimal control model.
Based on Kleinman et al. (1971).

$$\hat{\mathbf{X}}(t - \tau | t - \tau) = \hat{\mathbf{X}}(t - \tau | t - \tau - dt) + \mathbf{K}(t - \tau)$$
$$\times [\mathbf{Z}(t - \tau) - \mathbf{H}\hat{\mathbf{X}}(t - \tau | t - \tau - dt)], \qquad (3.32)$$

whereas the prediction portion of the task is performed using an optimal predictor,

$$\hat{\mathbf{X}}(t | t - \tau) = \Phi^i \hat{\mathbf{X}}(t - \tau | t - \tau), \qquad (3.33)$$

where $i = \tau/dt$. Finally the control portion of the task is performed using an optimal controller,

$$\mathbf{U} = \mathbf{C}(t)\hat{\mathbf{X}}(t | t - \tau). \qquad (3.34)$$

The optimal control model has been employed to describe several multi-input, multi-output manual control situations. The impact of the human having to control several variables is reflected in the covariance of the perceptual noise $\mathbf{V}_z(t)$. Based on the notion that the human must allocate attention (i.e., time) among the displays for each variable to be controlled, it is assumed that the variance of the perceptual noise for each variable is inversely proportional to the fraction of attention allocated to the display for that variable. Thus, as the number of controlled variables increases, performance degrades because the human must deal with increasingly noisy measurements.

Two other features of the optimal control model particularly deserve mention. Besides perceptual noise, the model also allows for perceptual thresholds. The combination of noise and thresholds allows one to use the optimal control model for evaluating displays, a topic we shall discuss shortly. The second feature of the optimal control model that should be noted is the assumption that the human has a perfect knowledge of Φ, Ψ, Γ, and so on. This is probably reasonably valid for highly trained operators of processes that respond quickly (e.g., automobiles and most aircraft). However, as the discussions in Chapter 2 indicated, it is unreasonable to expect that this assumption will hold in general. For example, processes such as jumbo jets, supertankers, and some chemical plants respond so slowly to control inputs that it is unlikely that the human could have a perfect knowledge of the process. This is partially due to there being very little proprioceptive feedback because slow processes do not allow the human to "feel" the results of control inputs. Thus, although the various noise sources in the optimal control model can at least partially handle situations where slightly less than perfect knowledge of Φ, Γ, and so on exists, there is a limit to the usefulness of this assumption.

Although it is still somewhat controversial (Dey 1975; Kleinman 1975; Phatak 1976; Baron 1977), the optimal control model appears capable of adequately representing a fairly wide range of manual control situations. Its main drawback is its mathematical complexity which can tend to obscure its internal workings and thus not permit the intuitive grasp possible with earlier models. However, this is mainly due to the complex multi-input, multi-output tasks that the model addresses. Several interesting applications of the optimal control model are reviewed in the August and October 1977 issues of *Human Factors* (Rouse 1977c). In the next section, we shall briefly discuss two papers dealing with display design that appeared in these issues.

Display Design Procedures

Curry et al. (1977) as well as Baron and Levison (1977) have employed the observation matrix **H** as well as the perceptual noise and perceptual threshold features of the optimal control model to consider display design questions on three levels of specification.

1. Information level—What will be displayed?
2. Display element level—How will it be displayed?
3. Display format level—How will it be integrated?

One can consider the information level using the **H** matrix of (3.21). If $h_{ij} > 0$, then state variable j is included in displayed output i. Otherwise, $h_{ij} = 0$. Thus, if $h_{ij} > 0$ for any i, state variable i is included on the overall display. Using **H** in this way, one can investigate the effects of adding and/ or deleting different state variables. For example, we shall later discuss work by Johannsen and Govindaraj (1980) where they used **H** to study the effect of adding predictions of state variables (i.e., a predictor display). In general, adding variables helps the human by eliminating the problem of inferring the values of variables that are not displayed. On the other hand, increasing the number of displays results in a decrease of the fraction of attention allocated to each display, and thus in an increase in the level of perceptual noise. Therefore, a trade-off emerges whereby one has to compare the advantages and disadvantages of various choices of which state variables should be displayed.

On the display element level, one has to choose among various devices. These devices can be characterized by several attributes including analog vs digital, electromechanical vs electronic vs computer-generated, monochromatic vs color, and so on. The choice of display devices affects the perceptual thresholds and perhaps the levels of perceptual noise. Thus, assuming perceptual data is available, one can assess the effects of various available devices. Alternatively, one could determine the impact of hypothetical display devices and thereby justify the investment of effort into developing such devices (or vice versa).

Considering the display format level, one can again use **H** to represent how several state variables might be integrated into a single display. Although the use of **H** does not tell one the best way in which to integrate variables, it can indicate the impact of successful integration. In other words, the optimal control model can be used to evaluate, but not directly generate, display ideas. A particularly interesting aspect of this type of evaluation involves considering the impact of differences in scaling possible with individual and integrated displays.

Baron and Levison (1977) consider some general display design issues that can be addressed using the optimal control model:

Is status information acceptable?

Will additional information degrade performance due to interference and/or high work load?

Do the advantages of display integration outweigh the improved scaling possible with separate displays?

Does command information integrate status effectively and, if not, how should it be done?

What performance and work load levels can be achieved with a perfectly integrated and scaled display?

Will quickening, prediction, or preview displays improve performance? What format should such displays have?

Many of the above issues[3] are considered by Baron and Levison in the context of designing an approach display for a Boeing 737. Curry et al.

---

[3] We have taken the liberty of adding the last item to Baron and Levison's list, mainly because we shall address that issue in the next section.

develop a 10-step display design procedure and apply it to a helicopter design study. Both of these applications are successful in illustrating the use of the optimal control model as an analysis tool, especially in terms of determining the sensitivity of performance and work load to various display features.

Quickening, Prediction, and Preview

When we discussed the solution to the optimal control problem, namely $U(t) = CX(t)$, we mentioned that it was particularly important to note that optimality required feedback of the complete state vector $X(t)$. Thus, for example, for a human to control a second-order system optimally, both position and velocity information are necessary. Similarly, control of a third-order system would require position, velocity, and acceleration information. In general, as the order of the system increases, more information must be available either via direct displays or by the human estimating, for example, the velocity as the rate of change of the position.

Unfortunately, human operators are not very good at estimating accelerations and higher-order derivatives. To overcome this difficulty, Birmingham and Taylor (1954) suggested that displays be "quickened" in the following way. Consider the output equation

$$Z(t + 1) = HX(t + 1) + V(t) \qquad (3.35)$$

for the case of a single-output system. In this situation, $H = [h_1 \, h_2 \cdots h_n]$. If one only displays the output position, then $H = [1\, 0\, 0 \cdots 0]$. With such a display, if n is large, humans will have difficulty controlling the system. However, if $h_1, h_2, \ldots, h_n$ are all appropriately nonzero, then the display will include information on all of the states of the system.

In a variety of situations, it has been shown that quickening the display by using $z(t + 1) = h_1 x_1(t + 1) + \cdots + h_n x_n(t + 1)$ with appropriate coefficients leads to improved manual control performance. The main difficulty with this type of display is that it is a "lie" in the sense that $z(t)$ is a weighted sum of position, velocity, acceleration, and so on and therefore ceases to have any unique physical meaning. For this reason, it sometimes is necessary to provide a conventional unquickened display, in addition to the quickened display, to provide the human with a true point of reference (Sheridan and Ferrell 1974, p. 270).

It seems reasonable to conclude that higher-order derivative information improves the human's performance by allowing the anticipation of future system outputs (i.e., prediction). If such predictions allow the human to perform better, perhaps performance could be improved even further by directly displaying predictions of future outputs to the human? In fact, *predictor displays* have been tried in numerous domains and substantial improvements in performance have resulted.

Predictor displays can be generated by using either a fast-time model of the controlled process or, by using a Taylor series extrapolation (Sheridan and Ferrell 1974, pp. 270–273). We shall consider the use of a Taylor series extrapolation as implemented by Johannsen and Govindaraj (1980) in a study of the use of predictor displays for longitudinal control in a VTOL hover task. In this application, predictor displays for both longitudinal position and pitch were considered. The prediction for longitudinal position x_1 was generated using

$$\hat{x}_1(t + T) = x_1(t) + Tx_2(t) + \tfrac{1}{2}T^2 x_3(t), \tag{3.36}$$

where x_1 = position, x_2 = velocity, and x_3 = acceleration, and T is the number of time units into the future being predicted. The predictor for pitch was similar in structure.

Using (3.36), predictions for three points in the future were generated: $\tfrac{1}{3}T$, $\tfrac{2}{3}T$, and T, where T equaled 2 seconds. Based on these three points, the curved path shown in Figure 3.3 could be approximated by the three straight line segments shown in the figure.

This type of display can be incorporated in the **H** matrix of (3.35) to yield

$$\begin{bmatrix} \hat{x}_1(t + \tfrac{1}{3}T) \\ \hat{x}_1(t + \tfrac{2}{3}T) \\ \hat{x}_1(t + T) \end{bmatrix} = \begin{bmatrix} 1 & \tfrac{1}{3}T & \tfrac{1}{18}T^2 \\ 1 & \tfrac{2}{3}T & \tfrac{2}{9}T^2 \\ 1 & T & \tfrac{1}{2}T^2 \end{bmatrix} \begin{bmatrix} x_1(t) \\ x_2(t) \\ x_3(t) \end{bmatrix}. \tag{3.37}$$

A somewhat similar augmentation of the **H** matrix allowed incorporation of the pitch predictor.

Johannsen and Govindaraj used the optimal control model of the human operator to evaluate various combinations of conventional and predictor displays. One particular interesting variation involved deleting the two midpoints of the longitudinal predictor [i.e., $\hat{x}_1(t + \tfrac{1}{3}T)$ and $\hat{x}_1(t + \tfrac{2}{3}T)$]. The analysis with the optimal control model indicated that deleting these midpoints would have little effect. Because this conclusion was somewhat counterintuitive, Johannsen and Govindaraj experimentally evaluated the

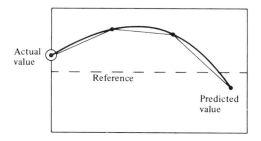

Actual
value

Reference

Predicted
value

Figure 3.3.

Johannsen and Govindaraj's longitudinal predictor.

Based on Johannsen and Govindaraj (1978).

idea and found that the midpoints were, in fact, not particularly useful to the human. This result is a good example of how the modeling approach can lead one down interesting research avenues that might not have been contemplated without the model.

We have already discussed how the optimal control model incorporates the fraction of attention devoted to each display. Namely, it is assumed that the variance of the observation noise for a display is inversely proportional to the fraction of attention devoted to that display. Kleinman (1976) has developed a method for determining the optimal allocation of attention to each display. It is interesting to consider whether or not the human optimally allocates attention. Using an oculometer to record eye movements, Johannsen and Govindaraj were able to compare empirical and optimal allocations of attention. Because the recording and interpretation of eye movements are fraught with difficulties, the results of the comparison were mixed. Nevertheless, the optimal fraction of attention for the longitudinal predictor display was found to agree quite closely with eye movement data for a helicopter pilot who served as one of the subjects in their experiment.

In all of our discussions of control thus far we have only considered the *regulation* problem in the sense that the human generates $U(t)$ in order to compensate for the disturbance $W(t)$ and thereby maintain the system's state as close as possible to $X(t) = 0$. This is called *compensatory* control. While many manual control models deal primarily with compensatory control, there are numerous situations where the control task is a *tracking* problem in that the human generates $U(t)$ in order to make $Z(t)$ follow some desired trajectory (e.g., the road in the case of automobile driving). This type of control in called *pursuit*. In fact, many realistic tasks can be viewed as a combination of tracking and regulation. For example, driving an automobile involves tracking the road while regulating against wind gusts. Thus, one can envision having a two-level model of manual control of automobiles (Donges 1978).

Consider a tracking task where the performance index of (3.13) can be modified to yield

$$J = E\left\{ \sum_{i=1}^{T} [Z(i) - Z_d(i)]'A[Z(i) - Z_d(i)] + U(i-1)'BU(i-1) \right\}, \quad (3.38)$$

where $Z_d(i)$ denotes the trajectory which the human is supposed to pursue or follow by generating $U(i)$. If, at time t, the human has knowledge of $Z_d(t+i), i = 1, 2, \ldots, T$, then the task is called *preview* control. Obviously, preview should improve performance, as is quite well-illustrated by the way in which a foggy evening affects automobile driving.

Tomizuka has developed a variety of manual control models for preview control tasks (Tomizuka 1973; Tomizuka and Whitney 1975, 1976; Tomizuka and Tam 1978; Tomizuka and Fujimura 1979). Using an optimal control formulation with the rate of change of U, rather than U itself, appearing in (3.38), Tomizuka found a control law of the form

$$U(t) = C(t)\hat{X}(t|t) + \sum_{i=1}^{T} C_d(t + i)\hat{X}_d(t + i|t + i), \qquad (3.39)$$

where $\hat{X}_d$ is generated from Z_d using a Kalman filter and, as might be expected on the basis of our earlier solution of the optimal control problem, the computation of C_d is somewhat complicated.

Tomizuka's experimental evaluation of the model showed that mean-squared tracking errors and control signal amplitudes were similar for model and subjects. Further, he found that the optimal preview length (i.e., the value of T necessary to achieve the best performance) was roughly the same for both model and subjects. In recent work, Tomizuka combined his preview control notions with the idea of quickening to yield a quickened preview display (Tomizuka and Tam 1978; Tomizuka and Fujimura 1979).

Govindaraj and Rouse have also studied preview control in the context of map displays for flight management (Govindaraj and Rouse 1979; Govindaraj 1979). More specifically, they were concerned with how a pilot schedules discrete tasks (e.g., talking with air traffic control, taking radio fixes, and so on) while also performing a continuous control task. This task presents difficulties for the usual optimal control models because it is typical for the pilot to stop controlling (i.e., stop moving the control stick) while performing some types of discrete tasks. As most manual control models produce continuous outputs, a new formulation was needed.

Using optimal control theory, Govindaraj and Rouse solved the preview problem for the case of deterministic Z_d. Their solution is similar to (3.39). The gains C and C_d were found to be inversely related to the weighting on control B in (3.38). This result led them to formulate the problem of scheduling discrete tasks as a problem of scheduling large values of B which would result in $C = C_d \approx 0$. Thus, they allowed B to vary over the planning horizon. Optimal scheduling of discrete tasks thereby was reduced to finding the optimal placement of large values of B.

It is important to note how this approach to achieving discontinuous control is quite different than simply setting $U(t) = 0$ for some period of time. The optimal control theory approach not only results in $U(t) = 0$ due to $C = C_d \approx 0$, where B is large, but also determines how $U(t)$, in the intervals for which it is nonzero, must compensate for the fact that a period of $U(t) = 0$ is approaching or has occurred. This notion of compensating for the fact that one cannot devote one's full attention to a particular task is important when considering multitask situations. This leads us on to the topic of supervisory control.

Supervisory Control

When a process is semiautomated or responds very slowly, it is not necessary for a human to devote full attention to that process, at least in the sense that control actions need not continuously be produced. This makes it possible for the human to be responsible for multiple processes. In situations where each individual process is automatically controlled, we can view the human as a supervisor whose role includes monitoring the processes to ensure that the automatic controls are working, adjusting the reference points (i.e., setpoints) with respect to which the automatic controllers are regulating the processes, and intervening in the case of failures and emergencies.

Sheridan and his colleagues have been studying supervisory control situations for quite some time (Ferrell and Sheridan 1967; Sheridan 1970; Sheridan and Ferrell 1974, Chap. 22; Sheridan 1976). Sheridan's view of supervisory control is quite general. However, in this chapter, we shall only consider those aspects of supervisory control that can be addressed with control theory. The more general aspects of supervisory control will be considered in Chapter 7.

Thus, at this point, we are interested in models of human decision making in the task of monitoring automatically controlled processes and, when necessary, intervening and adjusting the reference points. One of the most extensive studies of this type of modeling has been conducted by Kok and Van Wijk (1977, 1978). Their overall model is summarized in Figure 3.4. The supervised system is assumed to be automatically controlled. The automatic controls regulate the system with respect to the setpoints chosen by the human.

The model of the human supervisor is basically an extended optimal control model. Without doubt, the most significant extension is the decision-making function which includes thresholds that determine when setpoints will be changed and when observations, with which a cost is associated, will be taken.

Kok and Van Wijk's thesis, which is mainly theoretical in nature, presents derivations of a wide variety of properties of the model shown in Figure 3.4. They also compare the behavior of their model, in terms of the timing and amplitude of setpoint changes, to that of subjects in experimentation involving supervison of several slowly responding systems, including a supertanker.

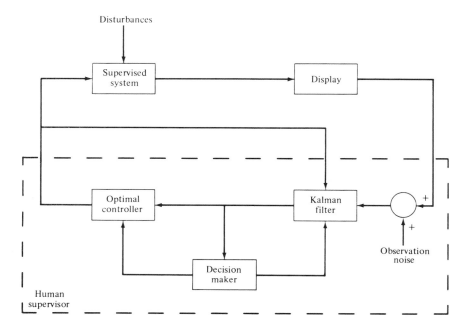

Figure 3.4. Kok and Van Wijk's model of supervisory control.
Based on Kok and Wijk (1977).

Muralidharan and Baron (1979) have studied supervisory control in the task of supervising multiple remotely piloted vehicles (RPVs). In this situation, the human has to choose which RPV to monitor and whether or not to intervene with discrete corrective control actions. Muralidharan and Baron's model incorporates the conceptual basis of the optimal control model while also including decision-making components. They have employed this model to study how uncertainties and the number of RPVs per operator affect overall system performance in terms of timing errors and deviations from the desired trajectory.

As the reader may have surmised from the publication dates of the above references, although the notion of supervisory control has been around for a while, control theorists have only recently been able seriously to approach the problem of developing control theoretic representations of supervisory situations. Further, since these efforts are so recent, little experimental data is available. Nevertheless, it appears that the human is becoming more and more of a supervisor (Sheridan and Johannsen 1976), and thus research will certainly continue in this area for quite some time.

Summary

In this chapter, we have briefly reviewed basic concepts of control theory with emphasis on discrete-time "modern" control formulations. This type of presentation has allowed us to avoid calculus, differential equations, Laplace and Fourier transforms, and so on. However, there is a cost to avoiding these topics. Namely, the rich history of manual control models in the "classical" control framework (e.g., the crossover model) has been quite tersely presented. Further, frequency domain analysis, which was briefly mentioned in Chapter 2, has received no discussion. Thus, this chapter presents a rather biased view of control theory, especially biased towards more recent work. If the reader is interested in a much fuller historical perspective and has the necessary mathematical prerequisites, then the books by Sheridan and Ferrell (1974) and Johannsen and his colleagues (1977) provide excellent presentations. In fact, Sheridan and Ferrell devote nine chapters to control theory models and all of the Johannsen et al. book is devoted to control theory.

However, although this chapter has slighted classical control theory and the frequency domain, we have included a reasonably thorough review of the state of the art in optimal control theory formulations of manual control tasks. We considered the problem of identifying the parameters of models of human control behavior. Our coverage of display design in general and quickening, prediction, and preview in particular should have amply illustrated the versatility of optimal control formulations in the sense that a wide variety of interesting manual control situations can be represented. Several other applications (e.g., car driving, assessing pilot opinion ratings, and anti-aircraft artillery tasks) are discussed in the previously mentioned special issues of *Human Factors* (Rouse 1977c).

We also briefly considered how optimal control formulations have recently been applied to modeling supervisory control situations. This work represents efforts to cope with the way in which the human operator's role is changing in the face of increasing automation. Although these new formulations look promising, one is inclined to wonder if the servomechanism analogy, and hence control theory, may not be approaching the limits of its applicability. We shall return to discuss this issue in Chapter 7.

Chapter 4
Queueing Theory

Estimation and control theory are concerned with system performance as defined by deviations of the system's state from some desired trajectory. For many interesting situations, performance can also be defined in terms of the length of time required to perform tasks. This performance measure can include the time that tasks must wait for attention as well as the time needed to actually perform the tasks. If one is concerned with waiting lines of tasks, then queueing theory may be the appropriate formulation. Within this chapter, we shall first consider the fundamentals of queueing theory based on books by White et al. (1975) and Kleinrock (1975, 1976) as well as a review by Allen (1975).

In contrasting queueing theory with estimation and control theory, a very interesting comparison is that of the analogies that each approach employs. In Chapter 2, we discussed how estimation theory models the human as an ideal observer whereas, in Chapter 3, we considered control theory models which represent the human as a servomechanism—an error nulling device. Queueing theory models of human–machine systems view the human as a time-shared computer, allocating attention and resources among a variety of tasks. In many ways, the time-shared computing analogy is more powerful than the ideal observer or servomechanism analogies, because it can represent a broader range of human–machine systems. On the other hand, when they are applicable, the ideal observer and servomechanism analogies allow significantly higher degrees of quantitative specificity. In Chapter 7, we shall further compare these different analogies.

Characterizing Queueing Systems

We shall describe the characteristics of queueing systems in the context of the multiple task human–machine system depicted in Figure 4.1. In this figure, we see that the human is responsible for N tasks and, at the moment, is attending to task 2. This figure can be considered as a human driving an automobile where the tasks include lateral control (steering), longitudinal control (accelerating/decelerating), scanning instruments, and so on. Another example is the aircraft pilot whose tasks include lateral control, longitudinal control, communicating with air traffic control, checking and updating radio fixes, monitoring the aircraft's subsystems, and so on. Yet another example is the human who monitors multiple processes in a power plant or a chemical plant. In this case, each task can be viewed as a particular process within the plant that potentially places a demand on the human's attention.

In most situations, not all of the tasks require attention simultaneously. In other words, tasks "arrive" at different times. One important characteristic of queueing systems is the *interarrival time distribution.* Many tasks, such as emergencies, arrive randomly. Such arrivals are said to follow a Poisson process which implies that the interarrival time distribution is exponential. A very important property of the Poisson process is the fact that the arrival time of the next task (or customer) is independent of when the last task arrived. Thus, it is said that a Poisson process has no memory. This lack of memory is called a Markov property. We shall make extensive use of this property in later discussions.

When one first encounters the idea of a Poisson process, a typical intuitive reaction is to note that very few tasks truly arrive randomly. Many tasks, for example, those of the air traffic controller, are actually scheduled. Thus, it would seem that deterministic (i.e., nonrandom) arrivals would frequently be an appropriate representation. However, if the schedule is very loose or is such that the schedule of each task is independent of the

Figure 4.1.
Multiple task human–
machine system.

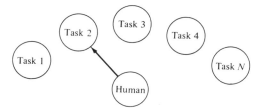

schedule of other tasks, then the flow of arrivals is quite likely to appear random.

The time required to perform each task is characterized by the *service time distribution*. There are a wide variety of possibilities. For many repetitive tasks, the service time is constant. However, when the task is less repetitive in the sense of not always requiring the same sequence of operations, then the service time will have some variability. If the amount of time required for completion of a task is independent of how long that task has already been in service, then an exponential distribution of service times is implied. Here again we see a process with no memory, and thus the process has a Markov property.

Although it is usually relatively easy to justify assuming Poisson arrivals, it is often difficult to show that an assumption of exponential service times is warranted. This is unfortunate because, as we shall see later, formulations with both exponential interarrival times and exponential service times are readily amenable to analytical solutions. Nevertheless, our models should reflect reality rather than analytical convenience. White et al. (1975, pp. 343–344) provide a useful rule of thumb for choosing a candidate service time distribution. It employs the mean and standard deviation as obtained from a sample of service time data. If the sample standard deviation is significantly less than the sample mean, an Erlang distribution should be considered. If the sample standard deviation is significantly greater than the sample mean, a reasonable candidate service time distribution is the hyperexponential. Finally, if the sample standard deviation and the sample mean are approximately equal, the exponential distribution is a good candidate.

Another important characteristic of queueing systems is the *service discipline*, which specifies the order in which tasks are performed. First come–first served (FCFS) is frequently employed. This is a typical service discipline when the task involves serving people who wait in line for various services. However, when people are not involved, in the sense that "fairness" from a social point of view need not be considered, several other disciplines are employed. Last come–first served (LCFS) often occurs when the arrangement of the waiting line is such that it is difficult to extract the first come customer. For example, if the task involves removing packages from a truck, one usually first removes those packages that were put in last.

Priority service disciplines occur when some tasks are more important than others. For example, an aircraft pilot would most likely assign a higher priority to an engine fire than to routine communications with air traffic control. Priority disciplines may be preemptive or nonpreemptive. A preemptive priority is such that the arrival of a higher priority customer causes the service of a lower priority customer to be interrupted. One interesting characteristic of preemptive priority disciplines is concerned with what happens after the preemption is complete. If task A preempts the servicing of task B and then, after the completion of A, servicing of B continues where it was left off, the discipline is called preemptive resume. However, if B must start all over again, the discipline is called preemptive repeat.

Other characteristics of queueing systems include number of parallel servers, maximum number of tasks allowed in the queueing system, and size of the task population. Servers are parallel when either of two or more servers could perform any particular task. Thus, for example, the pilot and copilot of an aircraft are, for the most part, parallel servers. However, the flight engineer is not an additional parallel server because his duties are quite different.

The number of tasks allowed in a queueing system (those in service plus those waiting for service) is often limited, especially because of space constraints. For example, the amount of space on a cathode-ray tube (crt) display is limited and thus only a finite number of tasks can be displayed. The notion of limited space raises the issue of what happens to tasks that arrive and find the system full. Queueing formulations usually assume that these tasks disappear and remain uncompleted. Although this type of behavior might seem quite unrealistic, it provides a useful means of exploring the effects of placing excessive demands on a system. Namely, the fraction of tasks that are not completed can serve as one index for assessing the viability of candidate system designs. This would allow determination of the amount of space and number of servers required to avoid uncompleted tasks.

The size of the task population simply denotes the potential number of tasks to be performed. Although in reality this number is always finite, it is analytically useful to assume an infinite population. Thus, whenever the population is large, one usually assumes an infinite population. For example, in an air traffic control situation, one could reasonably assume an infinite population of airplanes. However, within a single airplane, one would probably employ a finite population model to represent instrument scanning behavior.

Standard Queueing Notation

In order to easily denote the characteristics of queueing systems, a compact notation has been developed. Knowledge of this notation is particularly important when using the queueing theory literature because it frequently appears in the titles of papers and further, authors typically use it throughout papers without defining it. Queueing systems are classified using

$$(X/Y/Z):(U/V/W).$$

X refers to the interarrival time distribution, and Y denotes the service time distribution. Typical distributions are

$$M = \text{exponential } (M \text{ denotes the Markov property})$$
$$E_k = \text{Erlang}$$
$$HE_k = \text{hyperexponential}$$
$$D = \text{deterministic}$$
$$G = \text{general .}$$

As noted earlier, usually $X = M$ and Y varies considerably.

U refers to the service discipline. Typical examples include

$$\text{FCFS} = \text{first come–first served}$$
$$\text{LCFS} = \text{last come–first served}$$
$$\text{NPRP} = \text{nonpreemptive priority}$$
$$\text{GD} = \text{general .}$$

Z, V, and W denote the number of parallel servers, maximum number of tasks allowed in the system, and size of task population, respectively. Usually $V \leqslant W$ because it does not make sense to provide space for more than the maximum number of potential tasks. At least, it does not make sense to include the extra space in the analysis.

To illustrate the use of the standard notation, $(M/M/1){:}(\text{GD}/\infty/\infty)$ refers to a queueing system with Poisson arrivals, exponential service, a single server, general service discipline, infinite space within the system, and infinite task population. The most general queueing system that we can represent is $(G/G/c){:}(\text{GD}/N/N)$, which is such that all other queueing systems are special cases of it. Thus, if we could solve this queueing problem, we could solve all others. However, before we consider solving queueing problems, we must discuss performance measures.

Performance Measures

Queueing theory is concerned with modeling the use of two resources: space and time. From a human–machine systems perspective, space is of interest when designing displays, determining appropriate sizes for computer memories, and so on. Time is of interest because it is usually better to perform tasks quickly rather than have them wait. Also, one is concerned with time in the sense that the human's time is constrained in having only 100% to allocate.

Typical measures of queueing system performance include

L = expected number of tasks in the system

L_q = expected number of tasks waiting

W = expected time spent in the system

W_q = expected time spent waiting

D = probability of delay

ρ = server occupancy or utilization .

The measures with the subscripted q (L_q and W_q) represent the investment of resources in nonproductive activities (i.e., waiting). D is the probability of having to wait rather than immediately being serviced. One would usually like to have L_q, W_q, and D be as small as possible. This can be accomplished by having many parallel servers, the extreme case being one server for each potential task. However, this is expensive in two ways. First, servers usually must be paid. Second, having a large number of servers usually results in each server not being very busy (low server occupancy or utilization) which can lead to vigilance and motivation problems. Thus, one must trade off keeping ρ high (to save money and keep servers busy) and keeping L_q, W_q, and D low.

It is important to note that all of the queueing theory performance measures proposed here represent expected values of steady-state behavior. Although we also employed steady-state characterizations for estimation theory and control theory, it was a matter of convenience in those situations. For queueing theory, we shall see that it is more a matter of necessity.

Solving Queueing Problems

Given a particular queueing system of interest [e.g., $(M/M/c):(GD/N/\infty)$], solving a queueing problem involves determining L, L_q, W, W_q, D, and ρ. In this section, we shall illustrate how this is typically done.

We begin by defining the *state* of a queueing system. Quite simply, the state is the number of tasks n in the system. This includes both tasks being serviced and those waiting. The first step in solving a queueing problem involves determining $P_n(t)$, which is the probability of the system being in state n at time t. Consider $P_n(t + dt)$, where dt is assumed to be small. We are concerned with whether or not any new arrivals or service completions are likely to occur during dt.

To proceed further, we have to specify the queueing system of interest. We shall consider $(M/M/c):(GD/N/\infty)$. Thus, we are concerned with Poisson arrivals and exponential services. Let $1/\lambda_n$ and $1/\mu_n$ denote the mean time between arrivals and mean service time,[1] respectively, when the system is in state n. Since interarrival times and service times are assumed to be exponentially distributed, the means completely specify the distributions.

Notice that the subscripts on λ and μ allow for the possibility of arrival and service rates being dependent on the state of the system. This enables us to represent situations such as, for example, the concatenation of failures occuring in a machine once one thing fails. In this case, λ_n would increase, perhaps dramatically, as n increased.

The Markov property of the exponential distribution (i.e., its lack of memory) allows us to note the following:

$$\lambda_n \, dt = \text{probability of one arrival in } dt$$

$$1 - \lambda_n \, dt = \text{probability of no arrival in } dt$$

$$\mu_n \, dt = \text{probability of one service completion in } dt$$

$$1 - \mu_n \, dt = \text{probability of no service completion in } dt.$$

Assuming dt to be sufficiently small enables us to conclude that the probability of two or more events in dt (i.e., two arrivals, an arrival and a service, or two services) is approximately zero. Notice that the probabilities of arrivals and service completions do not depend on when the last arrival or service completion occurred. This illustrates how the Markov property is employed.

Using the above probabilities, we can write the basic *balance equation* for the $(M/M/c):(GD/N/\infty)$ queue.

$$P_n(t + dt) = P_n(t)(1 - \lambda_n \, dt)(1 - \mu_n \, dt) + P_{n-1}(t)\lambda_{n-1} \, dt(1 - \mu_{n-1} \, dt)$$

$$+ P_{n+1}(t)(1 - \lambda_{n+1} \, dt)\mu_{n+1} \, dt. \tag{4.1}$$

Thus, we see that the system can be in state n at time $t + dt$ if either (1) it started in state n and nothing happened, (2) it started in state $n - 1$ and one arrival occurred, or (3) it started in state $n + 1$ and one service completion occurred.

Because of the complexity (i.e., analytical intractability) of most queueing problems, we usually limit our interest to steady-state behavior. We can obtain the steady-state balance equation by solving (4.1) for $[P_n(t + dt) - P_n(t)]/dt$ and noting that this quantity must equal zero in steady state.

---

[1] We shall refer to λ and μ as the arrival *rate* and service *rate*, respectively. Thus, for example, instead of noting the average time between customer arrivals, we shall often mention the average number of customers who arrive per unit time (e.g., per minute).

This leads to (4.1) becoming

$$0 = -P_n(\lambda_n + \mu_n) + P_{n-1}\lambda_{n-1} + P_{n+1}\mu_{n+1}, \qquad 0 < n < N, \quad (4.2)$$

$$0 = -P_0\lambda_0 + P_1\mu_1, \qquad (4.3)$$

$$0 = -P_N\mu_N + P_{N-1}\lambda_{N-1}, \qquad (4.4)$$

where the last two equations reflect the fact that $\mu_0 = 0$ and $\lambda_N = 0$, respectively.

It is quite straightforward to solve (4.2)–(4.4) (White et al. 1975, pp. 92–93) to obtain

$$P_0 = \left(1 + \sum_{n=1}^{N} \prod_{k=1}^{n} \frac{\lambda_{k-1}}{\mu_k}\right)^{-1}, \qquad (4.5)$$

$$P_n = P_0 \prod_{k=1}^{n} \frac{\lambda_{k-1}}{\mu_k}. \qquad (4.6)$$

For special cases where $\lambda_n = \lambda$, $\mu_n = \mu$, and $N = \infty$, these equations become

$$P_0 = \left(\frac{(c\rho)^c}{c!(1-\rho)} + \sum_{n=0}^{c-1} \frac{(c\rho)^n}{n!}\right)^{-1}, \qquad (4.7)$$

$$P_n = \begin{cases} [(c\rho)^n/n!]P_0, & 0 \le n < c \\ (c^c\rho^n/c!)P_0, & n \ge c, \end{cases} \qquad (4.8)$$

where $\rho = \lambda/c\mu$ is the server occupancy which must satisfy $0 \le \rho < 1$.

With P_n determined, we can readily obtain

$$L = \sum_{n=0}^{\infty} nP_n, \qquad (4.9)$$

$$L_q = \sum_{n=c}^{\infty} (n-c)P_n. \qquad (4.10)$$

The waiting time characteristics W and W_q can be calculated using Little's formula (White et al. 1975, pp. 95–96)

$$L = \lambda W, \qquad (4.11)$$

$$L_q = \lambda W_q, \qquad (4.12)$$

which hold for a wide variety of queues and not just $(M/M/c){:}(GD/\infty/\infty)$. Instead of using both (4.11) and (4.12), one can employ either equation and the obvious relationship

$$W = W_q + 1/\mu. \qquad (4.13)$$

Finally, D can be shown to be (White et al. 1975, p. 103)

$$D = (c\rho)^c P_0/c! (1 - \rho). \tag{4.14}$$

A few things are particularly important to note about the results presented here. First, (4.9)–(4.13) are fairly general and do not depend heavily on the particular queueing problem being solved. Thus, relationships such as $\rho = \lambda/c\mu$ and $L = \lambda W$ are fundamental to almost all of queueing theory. On the other hand, the formulation of the balance equations is heavily dependent on the particular queueing system of interest. Further, in many cases, it can be quite difficult to formulate the balance equations and equally difficult to solve them.

To obtain the steady-state probabilities given by (4.7)–(4.8), we noted that ρ must be such that $0 \leqslant \rho < 1$. It is important to note why this is true, especially since we shall consider server utilization as a rough measure of human work load in our discussion of applications later in this chapter.

The server utilization is the ratio of average demands for services (λ) to average resources available to provide services ($c\mu$). One might wonder why queueing systems should not be designed such that $\rho = 1$. Such a design would seem to be most efficient. However, this approach would only work if customers arrived exactly on schedule and service time was constant. If a customer were ever late, time would be lost that could never be made up. Furthermore, if a servicing finished early, the time saved would be of no use because the server would be left idle until the next customer arrived.

Thus, unless a perfect schedule of arrivals and servicing can be maintained, one cannot design the system for $\rho = 1$. (If one does, the waiting lines will slowly become infinitely long.) Further, as the variability of interarrival times and service times increase, one must decrease ρ even more if long (but finite) waiting lines are to be avoided. A generally useful rule of thumb is that ρ should be less than 0.7. Of course, as with many rules, this rule is not always valid.

While it is possible to analytically solve a wide variety of queueing problems [e.g., $(M/M/c)$:$(GD/N/N)$, $(M/E_k/1)$:$(GD/\infty/\infty)$, and $(M/G/1)$:$(GD/\infty/\infty)$], many problems are analytically intractable. In such situations, one has two alternatives: approximation and simulation. Approximations are discussed in detail by Kleinrock (1976, Chap. 2) and we shall present one of his basic results later in this chapter. Simulation is very popular (Pritsker 1974; Schriber 1974). However, as noted in Chapter 1, it has the drawbacks of being expensive, difficult to interact with, and difficult to validate. Thus, one should only resort to simulation solution when it is necessary.

Even if one has to use simulation, the rigor of queueing theory is still quite useful. If one can formulate the problem of interest as a "standard" queueing problem in the sense of fitting it into the $(X/Y/Z)$:$(U/V/W)$

format, then even though one cannot solve the problem analytically it is quite likely that special cases of the formulation will be analytically tractable. Thus, one can use the analytical solutions to those cases to validate that the simulation is working as planned. Although this procedure cannot guarantee that all errors will be found, it still is quite useful.

Another benefit of a rigorous formulation, even for a problem that must be solved with simulation, is that it may allow use of basic theorems that have been derived for general classes of problems. For example, later in this chapter we shall discuss the optimal control of queues. We shall show how basic theoretical results in this area were successfully extrapolated for use in a formulation that required simulation solution. This success can be directly attributed to having formulated a "standard" queueing problem in the first place.

Standard Solutions

One of the most general queueing systems for which an analytical solution exists is $(M/G/1){:}(GD/\infty/\infty)$. Since the solution procedure is not as straightforward as that presented earlier for $(M/M/c){:}(GD/N/\infty)$, we shall only present the results here. However, it is useful to note that the general service distribution (G) is what complicates the solution. This is mainly due to the non-Markovian nature of a system with general distributions. In particular, the difficulty is that the amount of time in service remaining for a customer can *not* be assumed independent of how long he has already been in service. The solution to this problem involves artfully embedding a Markov process in the system. White et al. (1975, pp. 246–252) discuss the derivation. The final results are

$$L = \rho + (\lambda^2 \sigma_s^2 + \rho^2)/2(1 - \rho), \qquad (4.15)$$

$$L_q = \lambda^2(\sigma_s^2 + 1/\mu^2)/2(1 - \rho), \qquad (4.16)$$

$$W = 1/\mu + \lambda(\sigma_s^2 + 1/\mu^2)/2(1 - \rho), \qquad (4.17)$$

$$W_q = \lambda(\sigma_s^2 + 1/\mu^2)/2(1 - \rho), \qquad (4.18)$$

$$D = \rho, \qquad (4.19)$$

where σ_s^2 is the variance of the service time, whereas $1/\mu$ and $1/\lambda$, as before, are the mean times between task arrivals and service completions, respectively, and $\rho = \lambda/\mu$. Notice that using either (4.15) or (4.16) combined with (4.11)–(4.13) could have provided the remainder of the results for equations (4.15)–(4.18) Thus, some of our earlier basic results are still valid.

Since (4.15)–(4.19) are for a general service distribution, many interesting problems will be special cases of these equations. For example, the solution for $(M/M/1){:}(GD/\infty/\infty)$ specified by (4.7)–(4.14) can be obtained by assuming an exponential service distribution and thus $\sigma_s^2 = 1/\mu^2$. Equations (4.15)–(4.19) are also useful for determining the effect of service time variability. Consider the effect on W_q of having constant service times ($\sigma_s^2 = 0$) versus exponential service times ($\sigma_s^2 = 1/\mu^2$). From (4.18), we can see that constant service time yields one-half the waiting time that results with exponential service time. This clearly illustrates the cost of variability in a queueing system.

When we venture beyond $(M/G/1):(\mathrm{GD}/\infty/\infty)$, we start to employ approximations (Kleinrock 1976, Chap. 2). For example, for $(G/G/1)$: $(\mathrm{FCFS}/\infty/\infty)$, a bound on W_q is given by

$$W_q \leqslant \lambda(\sigma_a^2 + \sigma_s^2)/2(1 - \rho), \tag{4.20}$$

where σ_a^2 is the variance of interarrival times and σ_s^2 is the variance of service times. Notice that using $\sigma_a^2 = 1/\lambda^2$ (Poisson arrivals) in (4.20) yields a result approaching that of (4.18) as λ approaches μ. In other words, as the server occupancy ρ becomes close to 1, (4.20) offers a solution rather than a bound. Thus, for lower levels of demand, (4.20) provides a conservative bound.

Priority Queues

Priorities can be assigned to tasks in a variety of ways. Priority measures related to arrival times include first come–first served (FCFS) and last come–first served (LCFS). Measures related to service times include shortest-task-first (STF) and longest-task-first (LTF). Priorities can also be assigned based on membership in some group. For example, the head-of-the-line (HOL) priority, which we shall discuss shortly, fits in this category.

As discussed earlier, priority queues can be either nonpreemptive (NPRP) or preemptive (PRP). A NPRP queue is such that tasks in the *waiting line* are ordered according to priority numbers. Thus, those tasks already in service are unaffected by this ordering. On the other hand, a PRP queue is such that tasks in the *system* are ordered according to priority numbers *at any time*. Therefore, a task in service may be forced to go back into the waiting line by a higher priority task. Preemptive repeat systems are such that preempted tasks must be started all over again once they are taken from the queue. Preemptive resume systems are such that preempted tasks can be started again at the point where they were preempted.

Consider a $(M/G/1):(\mathrm{NPRP}/\infty/\infty)$ queueing system with K classes of tasks that vary in terms of

c_k = cost per unit time of delaying a class k customer, and

$1/\mu_k$ = average service time for a class k customer .

If the goal is to assign priorities in order to minimize

$$W_q = c_1\,W_{q1} + c_2\,W_{q2} + \cdots + c_K\,W_{qK}, \tag{4.21}$$

then it has been shown (Kleinrock 1976, pp. 119–126) that priorities should be assigned in terms of decreasing $\mu_k\,c_k$. This is termed a head-of-the-line (HOL) priority.

It is interesting to consider why the arrival rate for each class of tasks (λ_k) does not affect priorities. For example, consider a situation where $K = 2$ and $c_1 \gg c_2$. Suppose there are no class 1 tasks waiting for service but there are class 2 tasks waiting. Perhaps one should wait for a class 1 arrival because $c_1 \gg c_2$? After all, if one starts servicing a class 2 task, the NPRP nature of the system will not allow switching to a class 1 arrival until the task in service is complete. Thus, perhaps one should wait a short time dt in hopes of a class 1 task arriving. If a class 1 task does arrive during dt, then one would naturally service it. However, what if there are no class 1 arrivals during dt? Then, because of the lack of memory of the Poisson process, one is faced with exactly the same decision problem. To be consistent, one must choose to wait dt again. In this way, one would continually wait for class 1 arrivals and never service class 2 tasks. Thus, W_q would be infinite, which clearly does not minimize (4.21). From these arguments, one can see why this head-of-the-line priority scheme should not consider arrival rates.

To consider time-dependent priorities, assume that the priority of the jth arrival in class k is given by

$$p_{jk}(t) = c_k(t - \tau_{jk})^r, \tag{4.22}$$

where τ_{jk} is the time when the task arrived. For a $(M/G/1){:}(\text{NPRP}/\infty/\infty)$ queue using such a priority structure, Kleinrock (1976, pp. 126–135) has shown the expected waiting time for class k tasks to be

$$W_{qk}^r = \frac{[W_{qo}/(1 - \rho)] - \displaystyle\sum_{i=1}^{k-1} \rho_i\,W_{qi}[1 - (c_i/c_k)^{1/r}]}{1 - \displaystyle\sum_{i=k+1}^{K} \rho_i[1 - (c_k/c_i)^{1/r}]}, \tag{4.23}$$

$$W_{qo} = \sum_{i=1}^{K} \frac{\lambda_i(\sigma_{s_i}^2 + 1/\mu_i^2)}{2}, \tag{4.24}$$

where $0 \leqslant c_1 \leqslant c_2 \leqslant \cdots \leqslant c_K$.

The use of r in (4.22) provides considerable flexibility. For example, with $r = 0$, one can show that (4.21) and (4.23) lead to the μc HOL policy discussed earlier. Assuming that ties in priority are broken using a FCFS policy, one can even manipulate $p_{jk}(t)$ to obtain an overall FCFS policy (Kleinrock 1976, pp. 133–134). Thus, various results are special cases of the solution using (4.23).

Applications

Queueing theory formulations are particularly applicable to situations where task completion time is of utmost importance. Thus, for the most part, tasks have to be discrete in the sense of only requiring a limited amount of time. In other words, for task completion time to be important, the task must be capable of being completed in the sense that one does not perform the same task for the whole time period of interest. As a contrast, if one were required to perform a particular task for exactly 1 hour, then it would be unreasonable to choose task completion time as the performance measure.

One way to ensure that any particular task does not command all of the time available is to consider multitask situations where the human must allocate attention among tasks. Adding differing priorities among tasks as well as different interarrival time and service time distributions results in a multitask situation particularly amenable to queueing theory formulations. One constraint, however, is that interarrival times and/or service times should be probabilistic. If both arrivals and services are deterministic, the problem becomes somewhat trivial.

Since queueing theory is time oriented, it is quite natural to use such a formulation to consider the amount of time required by each task. More appropriately, one can consider the fraction of time devoted to each task. Also of interest is the total fraction of time required by all tasks. The server occupancy or utilization ρ can serve us quite well here. Let $1/\lambda_i$ and $1/\mu_i$, $i = 1, 2, \ldots, N$ be the mean time between arrivals and the mean service time, respectively, for the ith task of a set of N tasks. Then, $\rho_i = \lambda_i/\mu_i$ is the fraction of time required by task i and

$$\rho = \rho_1 + \rho_2 + \cdots + \rho_N \qquad (4.25)$$

is the fraction of time required by all tasks. Clearly, as we discussed earlier, a human–machine system should be designed so that $0 \leqslant \rho \leqslant 1$ and, in fact, as we also noted earlier, values of ρ at either extreme are undesirable.

Thus, the server occupancy can be viewed as a measure of work load. Such a measure is consistent with the way in which traditional timeline analysis views work load. However, when work load is viewed from a broader perspective (Moray 1979), one must conclude that server occupancy is only a first-order approximation and, in some situations, completely inadequate. Nevertheless, with a few minor modifications, we shall later illustrate how this gross measure has been useful.

A Visual Sampling Model

Carbonell (1966) and Carbonell et al. (1968) have studied the instrument scanning behavior of aircraft pilots and developed a queueing model that predicts the fraction of time devoted to each instrument. The basic assumptions underlying this model include the following:

The instruments compete for the pilot's attention in the sense that looking at one instrument postpones the observation of others, which therefore form a queue.

The queue discipline is such that the pilot attempts to choose an instrument for observation such that the total risk involved in not observing the other instruments is minimized.

The risk for each instrument is given by a cost particular to that instrument times the probability that the displayed value may, while not being observed, exceed a threshold particular to that instrument.

The time involved in reading each instrument is constant and, should the instrument be out of tolerance, the pilot will exert control actions appropriate to bringing the displayed value within tolerance.

Carbonell's model can be denoted $(X/D/1):(NPRP/N/N)$ in standard queueing notation. D stands for deterministic (i.e., constant) service times, NPRP denotes a nonpreemptive priority scheme, and N is the number of instruments. The interarrival time distribution is denoted by X because Carbonell's definition of customers in terms of the exceedence of a risk threshold makes it difficult to determine if arrivals fit any standard distribution. Carbonell solves this model using simulation.

The model was compared to the performance of three pilots who flew simulator approaches to Boston's Logan Airport. Eye movements and model parameters were measured. Also, a questionnaire was used to determine risk parameters. The fractions of attention devoted to instruments for heading, airspeed, altitude, pitch, roll, rate of climb, localizer, and glide path were considered. It was found that the model and pilots compared quite well in terms of the fraction of attention allocated to each of these instruments.

A Monitoring Behavior Model

Senders and Posner (1976) have developed a queueing model for situations where the human is a "pure" monitor in that the human observes processes but does not control them. The research was specifically concerned with manning requirements and the reliability of human–machine interaction in terms of process failures due to limit exceedences being unobserved.

Senders and Posner modeled the class of monitoring situations of interest as a $(G/M/1):(\text{FCFS}/N/N)$ queue. Analytical solutions for the average delay W_q and probability of instruments exceeding acceptable limits were developed. Defining a process failure as a limit exceedence, they determined the expected number of process failures within a given time period of interest.

Although this model has not been compared to experimental data, it appears that it would be useful for evaluating the trade-off between number of operators and probability of process failure. At one extreme, the utilization of each operator (i.e., ρ) should not exceed one and, in fact, should probably be considerably smaller. On the other extreme, one operator could be assigned to each instrument. This would supposedly minimize the probability of limit exceedence and hence of process failure.[2] However, it would be quite costly and might result in vigilance and motivation problems. Thus, one might use Senders and Posner's model to trade off reliability and cost.

An Air Traffic Control Model

Schmidt (1978) employed a queueing theory formulation to analyze the work load of air traffic controllers. Analysis of field data from the Los Angeles Air Route Traffic Control Center led to describing the air traffic control task as a $(M/E_k/1):(\text{NPRP}/\infty/\infty)$ queue. This model was then used to predict average delay (W_q) and server occupancy (ρ) as a function of demand characterized by λ.

This type of formulation would seem to be useful in several ways. Most obviously, it could be used to make staffing decisions. In other words, the queueing analysis would help in scheduling the number of air traffic controllers needed during different periods of the day and days of the week. Further, during times of peak loads, it could be combined with a runway capacity model and used to estimate W_q and thereby provide information to incoming aircraft concerning delays. Finally, the work load component of the model (i.e., server occupancy) could be incorporated in a computer system that monitors the air traffic control system and provides assistance to the controller. We shall discuss this notion further later in this chapter.

---

[2] Unless, of course, assigning a single operator to each instrument results in such low utilization per operator that problems of underload, as opposed to overload, arise.

A Flight Management Model

In the remainder of this chapter, we shall discuss a particular use of queueing theory in some detail. Our goal in this discussion will be to clearly illustrate how and why queueing models can be useful.

Rouse and his colleagues (Rouse 1977b; Walden and Rouse 1978; Chu and Rouse 1979) have employed queueing theory to model pilot decision making in a multitask flight management situation. The scenario ran as follows. A PDP-11 driven crt graphic system was employed to represent a cockpit-like display to an experimental subject. The display shown in Figure 4.2 included standard aircraft instruments such as artificial horizon, altimeter, heading, and airspeed indicators. Also displayed was a flight map which indicated the airplane's position relative to the course to be followed. A small circle moved along the mapped course indicating the position the aircraft should have for it to be on schedule.

The aircraft could be flown in two modes: autopilot and manual. In the manual control mode, the pilot controlled the pitch and roll of Boeing 707 aircraft dynamics with a joystick. Another control stick regulated the airspeed. The pilot's control task was to fly the airplane along the mapped route while maintaining a fixed altitude and stable pitch and roll attitude.

Below the map were dials that represented the numerous aircraft subsystems which the pilot monitored for possible action-evoking events. Upon detecting an event (represented by the normally motionless pointer drifting to a downward postion as shown for the engine subsystem in Figure 4.2) to which they wished to respond, the subjects selected that subsystem via a 4×3 keyboard. The display shown in Figure 4.3 then appeared. This represented the first level of a checklist-like tree associated with the subsystem of interest. The subjects then searched for a branch labeled with a zero and selected that branch with the keyboard. The next level of the tree was then displayed, and so on. After completing the last level of the tree, the action was completed and the display shown in Figure 4.2 returned, with the subsystem information or diagnostic check complete.

The subsystem events were scheduled to arrive according to a Poisson distribution. Events of different subsystems arrived independently with fixed priority. The subjects were instructed to place a higher priority on the control tasks than on subsystem tasks; within subsystem tasks, priority decreased from left to right. For example, the navigation subsystem was the most important while the cabin temperature subsystem was the least.

Queueing theory was used to describe this multitask decision-making situation in the following manner. Since the monitoring task was, in fact, a queueing system, it was easily modeled as a single-server queueing system with subsystem events as customers. The control task was incorporated as a special queue. Customers arriving in the control task queue could preempt the service of a subsystem event. Service of the subsystem event would then be resumed at the point of interruption after the control action was

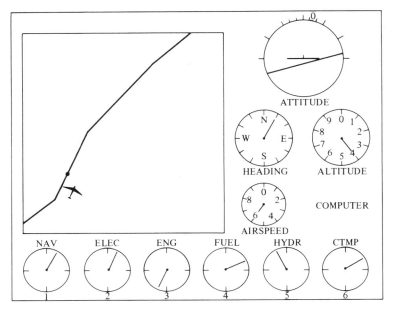

Figure 4.2. The flight management situation.

Based on Walden and Rouse (1978).

complete. This queueing system was described as a $(M/E_k/1)$:$(PRP/N/N)$ queue. The model was solved using simulation because an appropriate analytical solution was not available.

One of the problems with this model was defining a "customer" for the control task queue. In fact, it was displayed error (more specifically, increments of displayed error) which queued (accumulated) for attention. However, the size of the incremental "customer" was unknown. Assuming that control task customers preempted subsystem service, a customer in the control task queue could be defined (loosely) as a "significant" or "action-evoking" amount of displayed error. A simple way to measure indirectly the frequency with which "significant" errors arrived (queued) was to measure the frequency with which displayed errors were serviced (i.e., the frequency at which control actions were inserted to null them).

Thus, to estimate the frequency of control responses, the number of separate control actions performed by the pilot could be counted, and from this number an arrival rate for control task customers could be obtained.[3]

---

[3] If control actions were continuous rather than discrete, then a threshold would be necessary for defining control task customers. Such a threshold would be difficult to measure and thus would have to be either a free parameter or chosen on the basis of previous experiments.

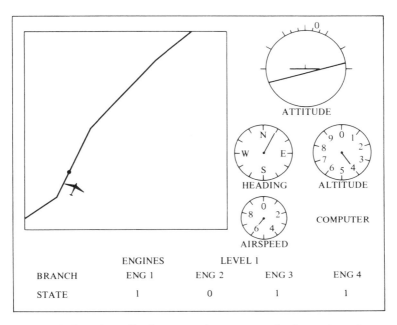

Figure 4.3. Display when pilot has reacted to an event in the engine subsystem. Based on Walden and Rouse (1978).

On the other hand, the average service rate for control task customers was more difficult to measure since it was not necessarily related to the length of time during which the controls were continuously nonzero. Thus it was left as a free parameter (the only one) of the model.

Using the experimental situation described above, two experiments were performed to study human performance in the multitask flight management situation. The two independent variables in the first experiment were the interarrival times of subsystem events and the difficulty of the flight path to be followed by the pilot. Four levels of subsystem event interarrival time were studied, including 30-, 60-, and 90-sec average interarrival times per subsystem as well as a case with no arrivals (i.e., without the subsystem monitoring task). Three levels of control task difficulty were investigated including no map (i.e., no control task), simple map with few turns, and complex map with many turns. Six subjects participated in this experiment, the design of which was balanced with respect to the average interarrival rate for the monitoring task.

Performance measurements during the experiment included:

time of occurrence of a subsystem event,

time of response to the event,

time of completion of checklist for the event,

aircraft position and attitude, and

pilot control inputs.

These measurements were used to determine input parameters for the $(M/E_k/1):(\text{PRP}/N/N)$ model. These parameters included:

probability of incorrect action, given that a subsystem event had occurred;

false alarm arrival rate and service rate;

subsystem event arrival rate λ_s, one of the independent variables in the experiment;

average subsystem event service rate μ_s, obtained from the empirical service time data;

Erlang subsystem service time distribution shape parameter k_s, the square of the ratio of mean to standard deviation (calculated from the empirical service time data and averaging 19);

control task arrival rate λ_c, obtained from the control histories;

control task service rate μ_c, the only free parameter in the model;

Erlang control task service time distribution shape parameter k_c (a value of 2 was used).

As noted earlier, the model was solved using simulation. For each combination of monitoring task difficulty and control task difficulty, 10,000 events were simulated. The output of the model and the empirical data were compared in terms of the average and standard deviation of the waiting time for events in each of the six subsystems. Thus, 36 comparisons were possible (6 subsystems × 3 subsystem event arrival rates × 2 statistics) for each level of control task difficulty. As noted earlier, control task service rate μ_c was left as a free parameter. It was varied to obtain the best fit possible between model and data. (Of course, in the "no map" situation, μ_c was irrelevant.) In general, the comparisons between model and data were quite favorable (Walden 1977; Walden and Rouse 1978). Now, let us consider how this model was applied to the evaluation of a potential new approach to human–computer interaction.

Modeling Human–Computer Interaction

Computers are increasingly being used in a number of decision-making situations, especially when several tasks have to be performed at the same time by a human decision maker. Commercial aircraft can now, in many situations, be flown from takeoff to landing using the autopilot and navigation systems. Industrial processes can be monitored and controlled by computers. However, because the performance and functional demands on the system are so great, it appears that the need for the human in the system, supervising and managing the operation, has not diminished. In such situations, the human has to interact with computers which are capable of processing and routing information, exerting control actions, and making choices in view of priority conflicts. An important issue that arises concerns the roles that human and computer should play as systems become increasingly automated. One very basic question is: How should decision-making responsibilities be allocated between humans and increasingly intelligent computers?

As a straightforward approach, one might allocate a fixed portion of the set of the tasks to the computer with the remainder of the set being allocated to the human. Licklider (1960) has proposed that the human set goals, formulate hypotheses, determine criteria, and evaluate results. On the other hand, the computer should perform routine work such as transforming data, simulating models, and implementing results for the human decision maker.

However, the division of tasks is not as clear cut for decision-making situations where computerized decision aiding systems are included. Rouse has suggested that a dynamic or adaptive allocation of responsibilities may be the best mode of human–computer interaction (1977b). With adaptive allocation, responsibility at any particular instant will go to the decision maker most able at that moment to perform the task. Such a scheme is adaptive in the sense that the allocation of responsibility depends on the state of the system as well as the states of the decision makers. Thus, changes in system or decision-maker states result in changes in the allocation policy in order to optimize performance.

The adaptive policy that will be discussed here allocates decision-making responsibility in order to optimize system performance subject to maintaining human work load at appropriate levels. Further, to avoid increased human work load due to having to make allocation decisions, the allocation decision is automated and delegated to a computerized coordinator. The coordinator *recommends* particular allocations of task responsibilities which the human may choose to preempt. Considering a task domain where the computer is employed as back-up decision maker, the problem is further simplified by assuming that the coordinator has all the information needed and that both human and computer have common goals. Then the simplified coordination problem involves answering the following question: When should the computer request and relinquish responsibility?

Rouse has described human–computer interaction in multitask decision-making situations as a queueing system with two servers (human and computer) and K classes of customers (1977b). With this description, the problem of allocating decision-making responsibility is simplifed to one of determining who serves a particular customer, or equivalently, to which server the arriving customer should be directed. Thus, we have a queueing control problem.

Using a queueing system framework, Markov decision process models have been employed by many researchers to represent queueing control problems. A thorough review of the literature with emphasis on the dynamic control of queues using service variables, arrival variables, and priority disciplines is given by Chu (1976). We shall only review two particularly important results here.

Heyman (1968) considered the problem of controlling a queueing system with Poisson arrivals, general service time distribution and single server $(M/G/1)$ by turning the service mechanism on when a customer arrives or off when a customer leaves. He showed that the optimal stationary policy which minimizes linear average or discounted cost over an infinite horizon has a simple critical number characterization: (S, s). This (S, s) policy is to provide no service if the system size L' (i.e., number of customers in the queue) is s or less, and to turn the server on when the size L' is greater than S. The cost incurred includes waiting cost, running cost, and switching cost.

Bell (1973) extended this result to an $M/G/1$ nonpreemptive priority queue and proved the existence of an optimal average cost policy of the (a, b, c) type for two priority classes. This optimal policy is the following: never turn the server off or turn the server on the first time that $an_1 + bn_2 \geqslant c$, where n_1 and n_2 are the number of class 1 customers and class 2 customers in the system. For the general situation of K priority classes, the optimal control actions are simply characterized by the $(K - 1)$-dimensional hyperplane of the form: $a_1 n_1 + a_2 n_2 + \cdots + a_K n_K = c$.

Using the above theoretical results, Chu and Rouse (Chu and Rouse 1979; Chu 1978) advocated the use of the stationary expected cost optimal policy for computer on/off of the following form: turn the computer on at arrival epochs when $L' = c_1 n_1 + c_2 n_2 + \cdots + c_K n_K \geqslant S$ and turn it off when $L' \leqslant s$, where $S, s, c_1, c_2, \ldots, c_K$ are nonnegative constants and $n_k = 0$ indicates that task k is not in the system (i.e., in service or waiting), whereas $n_k = 1$ indicates that task k is in the system. The c_k are chosen according to the relative priorities of events. Bell's results (1973) imply that the c_k here happen to be the same as the assigned constant cost rates c_k for single server, two priority process situations. The choice of this weighted threshold policy (which depends on the number of customers present) was based on the ease with which its inputs could be measured, its responsiveness to variations of arrival rate and service rate, and the fact that considerable literature suggests this measure.

The optimal threshold policy (i.e., S and s) should vary as the system characteristics vary. The sources of variation include (1) traffic demand (arrival rates), (2) server performance (service rates and probabilities of error), and (3) system and performance uncertainties (unidentified parameters). Chu and Rouse advocated implementing the adaptive optimal policy by setting up a table of stationary control policies off-line and then implementing a table look-up procedure along with on-line identification and estimation of system variables.

Although it is possible to derive an analytical procedure for determining the optimal thresholds (Chu 1978), it is very cumbersome and requires reformulation for each set of thresholds. Thus, a simulation approach was adopted to determine the optimal stationary policy. The simulation program mentioned during our earlier discussions of flight management was modified to include the computer aiding notions discussed above. The modified program represented a $(M/G/2):(\mathrm{GD}/N/N)$ queue, where use of the second server was controlled by the threshold policy considered earlier. Program validity was tested by comparing the values of W_q produced by the program with the analytical solution for a $(M/M/c):(\mathrm{GD}/N/N)$ queue, where both servers were "on" at all times. There were no statistically significant differences between simulation and analytical solutions for this special case.

To evaluate the weighted-threshold approach to computer-aiding, the flight management task discussed earlier was modified to include adaptive aiding. The computer was assumed to be able to perform monitoring and diagnostic check procedures using information from channels linked with subsystem computers and from the data links. It made no errors such as false alarms, missed events, or incorrect actions. The detection and service times were assumed constant. The computer employed the same priority rule among subsystems as that used by the pilot. To be consistent in its back-up role, the computer adapted itself to and avoided interference with the pilot. To this end, the pilot was allowed to override any recommendation that the computer offered.

To inform the pilot of the computer's action, a succinctly displayed computer status indicator on or near the subsystem displays appeared to be satisfactory. For example, in Figure 4.4, the "NAV" symbol over the navigation dial flashed if the computer decided that an event had occurred (and the threshold exceeded) and was waiting to be serviced in the navigation system. The purpose of this indicator was to inform the pilot that it was possible to take charge of the navigation system without any

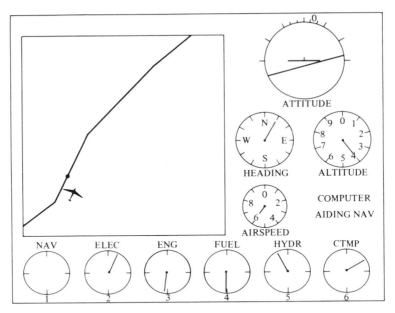

Figure 4.4. Display when computer is performing a check of the navigation subsystem.

Based on Chu and Rouse (1979).

computer interference. Otherwise, the symbol would continue to flash for a total period of 4 seconds until the computer started interacting with the navigation system, resulting in a dim indicator showing in the navigation dial. If the pilot was in the middle of performing some other subsystem check procedure, say within the engine subsystem, the flashing "NAV" symbol over the navigation dial would not be seen. The status of the computer was then shown on the lower right corner of the crt by an "AIDING NAV" symbol (flashing during the interval of possible pilot preemption), if the computer was awaiting preemption or interacting with the navigation subsystem. This computer status area was blank if the computer was not actively involved in the subsystems.

Pilot-to-computer communication, if viewed in general, is very complicated (Rouse 1977b). However, in this study, the communication channel from the pilot to subsystems was narrowly defined. These channels included the keyboard input and stick response sampling (through an analog-to-digital converter). These channels provided the monitoring computer a way of determining if the pilot was interacting with any portion of the system. If a number had been received through the keyboard, and the checklist was being processed, then the pilot had to be performing a subsystem task.

Deviation of the control stick from normal position revealed that the pilot was performing the control task.

To obtain the stationary policy (i.e., to determine the values of S and s) suitable for the experimental situation, a computer simulation was performed. Based on the results of Walden and Rouse (1978), the system was represented as a preemptive resume priority queueing system: $(M/E_k/2)$: $(PRP/N/N)$ with implemented threshold control. System parameters included the following:

Subsystem arrival rates and service rates which were all uniform among the subsystems. For waiting costs, $c_1 = c_2 = \cdots = c_K = 1$ was used.

Two levels of arrival rates were assumed, i.e., low arrival (at 0.0167 events per second per subsystem) and high arrival (at 0.0333 events per second per subsystem).

Pilot performance in terms of service rates, service errors and control services were obtained using the data from Walden's experiment.

The computer aiding automatically went off when no event needed service (i.e., $s = 0$) and when in service, the computer employed approximately the same service time (7 sec) as the pilot.[4]

The results based on the computer simulation of 10,000 events for $N = K + 1$ (i.e., 6 subsystems plus 1 control task) and a desired server occupancy for the pilot of $\rho \leqslant 0.7$ showed that, without the control task, $S \leqslant 7$ for low arrival and 3 for high arrival; with control task, $S \leqslant 3$ for low and 1 for high arrival. This choice of 0.7 as the server occupancy threshold was based on the aforementioned rule of thumb for simple queueing systems where a higher value of occupancy will result in a steep rise in average queue length. These threshold values are those which the computer employed to adapt to both subsystem arrival rate and control task involvement to minimize expected subsystem waiting time subject to the desired occupancy level. Obviously, for systems with different values of K, λ, μ, or ρ the appropriate threshold values are likely to be different from those listed above.

An experiment based on the representation described above was conducted. Eight thoroughly trained subjects, all of them male students in engineering, participated in a balanced sequence of 16 experimental runs with different work load levels. This was achieved by combining three levels of control task involvement (perfect autopilot, manual control, autopilot with possible malfunctions), three levels of subsystem event arrival rates (no

---

[4] By using the same service rate for both pilot and computer, it was possible to avoid confounding the availability of computer aiding with possible speed mismatches. Further, the state of the art is such that any computer program that would be reliable and sufficiently intelligent to perform the tasks discussed here would also be relatively slow. Discussion of the effects of speed mismatches can be found in Rouse (1977b).

arrival, low arrival, high arrival), and three levels of availability of computer aiding (no aiding, aiding with fixed switching policy, and aiding with adaptive policy) while excluding the degenerate cases such as no arrival and perfect autopilot, no arrival and adaptive aiding, and so on.

For the experimental runs with perfect autopilot, only the subsystem task was considered. An "autopilot" kept the aircraft on course and on schedule. These runs provided measures of baseline performance for the subsystem task. In the manual control runs, subjects had to perform both subsystem and control tasks. As noted earlier, they were told that the control task was more important than the subsystem task (i.e., control arrivals preempted subsystem arrivals). For the runs where autopilot malfunctions were possible, the autopilot was available during most of the experiment so that subjects were not required to fly the airplane except to occasionally check autopilot performance. As soon as they detected an autopilot malfunction, which was characterized by the airplane deviating from the mapped course at a turn rate of one degree per second, they were required to take over the flight control task and fly the airplane back to the mapped course, whereupon the autopilot mode was restored.

After detecting the autopilot malfunction, the pilot had to devote a major portion of attention to the control task, leaving subsystem tasks less attended. Thus, in this period, the pilot's work load suddenly increased. To adapt to this type of change, a lower threshold value was used to reduce subsystem service delay and pilot work load. Two experimental runs with adaptive computer aiding were included in the set of runs with autopilot malfunctions possible. Instead of using $S = 3$ all the time as in the fixed threshold policy the adaptive policy used $S = 1$ whenever the pilot was in manual mode.[5]

We can summarize the experimental results as follows: control involvements, subsystem event arrival rates, and the availability of computer aiding significantly affected subsystem waiting time and server ocupancy as well as subjective effort ratings. It was observed that systems that are designed to relax control requirements, such as the autopilot, seem to improve both control and subsystem performance, whereas systems, such as computer aiding, that are designed to relax subsystem requirements seem to improve only subsystem performance. The possible reason for this is that the control tasks preempted subsystem tasks, and thus control task inefficiency was likely to affect the performance of subsystem tasks, but if the assumed preemption rule holds the reverse would not be true.

Server occupancy and subjective effort ratings were highly correlated. Aiding was able to enhance system performance in terms of average

---

[5] It should be noted that adaptive policy means adjustments of the on/off threshold in response to changes in what the pilot is doing. The on/off strategy with fixed threshold is really not adaptive in the usual sense of the word.

subsystem waiting time, server occupancy, and subjective effort ratings. Adaptive aiding was shown to further reduce subsystem waiting time. Interestingly, adaptive aiding did not significantly improve service occupancy. However, it did improve system performance.

Now we shall consider the adequacy of the queueing model of the computer-aided flight management task. First, we have to discuss how various performance characteristics were included in the model. Human false alarms, human control actions, and autopilot malfunctions were considered to be separate processes with given arrival and service statistics and with appropriate interactions with each other. Features of computer aiding such as the preemption period were easily implemented. The following additional parameters were incorporated within the program:

> subsystem scanning time (0.25 sec per subsystem for human, 0.0 for computer), and
>
> monitor/control attention shift (0.2 sec).

In the simulation program, all process arrivals (including subsystem arrivals, false alarm arrivals, autopilot malfunction arrivals, and control action arrivals) were generated using a Poisson distribution, and all service times (including service of subsystem events, incorrect actions, false alarms, autopilot malfunctions, and control actions) were assumed to follow an Erlang-k distribution. In the cases of subsystem and false alarm services, the service time distributions were approximately constant (i.e., large k). The set of variables used in the program represent values measured from the experiment and averaged across all appropriate situations. These variables served as input to the program and included the following:

> the subsystem service time distribution (with mean of 5.668 sec and $k = 62$);
>
> the control service time distribution (with mean of 2.13 sec for manual mode, and 2.99 sec for malfunction mode, $k = 2$ for both);
>
> the false alarm arrival rates (with mean of 0.00344 arrivals per second for low arrival rate, 0.00915 arrivals per second for high arrival rate);
>
> the probabilities of incorrect actions (of 0.0656 for low arrival rate, and of 0.0865 for high arrival rate);
>
> the autopilot malfunction detection and service time distribution (with mean of 7.28 sec and 30.15 sec, respectively, and $k = 2$ for both);
>
> the incorrect action service time distribution (with mean of 3.50 sec and $k = 5$); and
>
> false alarm service time distribution (constant of 1.56 sec).

All parameters in the model were either predetermined or empirically measured with no adjustments made. Comparisons of the model average subsystem waiting time and model server occupancy with those measured from the experiment were quite favorable. In addition, high correlations ranging from 0.75 to 0.96 were found between the model's predictions of server occupancy and the effort ratings of individual subjects. Thus, since one of the best ways to measure work load is still to ask the human to rate his own work load, the model may be useful for predicting levels of work load in multitask situations.

Summary

In this chapter, we have introduced a time-shared computer analogy of human decision making in multitask situations. This led us quite naturally to the use of queueing theory for modeling human–machine systems where the time-shared computer analogy is appropriate. We proceeded to show how queueing problems are formulated and subsequently, considered how one can determine the performance characteristics of queueing systems.

In discussing applications, we noted that queueing theory is most appropriate when time is the dominant performance measure. Measures of particular interest include the time that tasks must wait and the fraction of time that servers are busy. We briefly considered server utilization as a measure of work load and mentioned the difficulties possibly resulting from either overload or underload. One application in particular illustrated the usefulness of this work load measure.

The five applications of queueing theory discussed in this chapter served to emphasize the fact that queueing theory models are most appropriate in multitask situations where the human must deal with priorities and performance requirements which vary among tasks. Thus, queueing theory models are concerned with how a human coordinates the performance of a set of tasks rather than how he performs a particular task. In other words, one might say that queueing theory is more concerned with *when* rather than *how* tasks are performed. From this perspective, one can see the possibility of modeling higher levels of a task hierarchy with queueing theory and lower levels of the hierarchy with, for example, estimation and control theory. We shall return to this issue in Chapter 7.

As a final point, we want to emphasize the usefulness of formulating problems in rigorous queueing theoretic terms rather than developing somewhat ad hoc simulation models. First of all, as noted earlier, rigorous formulation may allow one to employ various useful theorems available within queueing theory. Second, rigorous formulation may help one to test a simulation using special cases that are analytically tractable. Finally, attempting to use standard queueing theory formulations should help to provide a common fabric within which various models of human–machine interaction can be woven.

Chapter 5
Fuzzy Set Theory

Thus far, we have looked at the human as an ideal observer, a servomechanism, and a time-shared computer. In this chapter, the human will be viewed as a logical problem solver. This perspective will lead us to describe human problem solving behavior as sequences of logical set theory operations. We shall assume that the logical operations are unaffected by human limitations. However, the notion of constrained optimality will be incorporated in the overall model by allowing the sets on which operations are performed to be defined in an approximate or fuzzy manner. Thus, as the reader will see shortly, we shall be concerned with the use of nonfuzzy operations performed upon fuzzy sets.

The basic material on fuzzy set theory in this chapter is drawn from books by Zadeh et al. (1975) and Kaufman (1975), as well as a review by Tong (1977). As evidenced by the bibliography of Gaines and Kohout (1977), the literature of fuzzy sets is immense. This chapter will only scratch the surface, presenting sufficient detail to illustrate some applications of fuzzy set theory. It is important that the reader realize that an enormous amount of almost purely theoretical literature on fuzzy sets exists and that the field of study extends far beyond this chapter. Nevertheless, within this chapter, we do hope to provide a solid understanding of the primary and basic concepts of fuzzy set theory while also illustrating that these novel concepts are of considerable practical value.

Nonfuzzy Sets

It is perhaps useful to begin by first briefly reviewing a bit of nonfuzzy set theory. A set is composed of *elements*. The set of all elements is called the *universe* and is denoted by U. A *subset* of U, denoted perhaps by A, is such that all of the elements of A are also elements of U. Figure 5.1 illustrates such a relationship. Note here that an element of U either does or does not belong to A. Thus, for example, U might include all the people in the world while A includes only children under 10 years of age. Consider how one might define the elements of A if they were simply defined as all "young" children. In this situation, it is not quite clear what "young" means and therefore, it might be difficult to decide whether or not a particular child belongs in A. We shall return to this issue a bit later.

There are three basic set theory operations that will be useful throughout this chapter: union, intersection, and complementation. The *union* of two sets (or subsets) A and B, denoted $A \cup B$, is illustrated in Figure 5.2. The shaded area includes all elements that are in A, in B, or in both A and B. Thus, for example, $A \cup B$ could include all children that are less than 10 years old or under 25 kg in weight or both. A particular child is a member of $A \cup B$ in as much as *either* of these two conditions is satisfied.

The *intersection* of two sets A and B, denoted $A \cap B$, is depicted in Figure 5.3. The shaded area includes all elements that are in both A and B. Thus, continuing our example with children, $A \cap B$ would include all children that are less than 10 years old and under 25 kg in weight. A particular child is a member of $A \cap B$ to the extent that *both* of these conditions are satisfied.

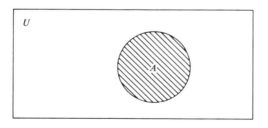

Figure 5.1.
Illustration of A as a subset of U.

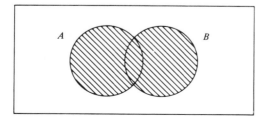

Figure 5.2.
Union of A and B.

Figure 5.3.
Intersection of *A* and *B*.

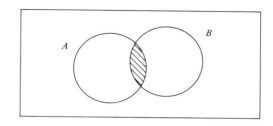

Figure 5.4.
Complement of *A*.

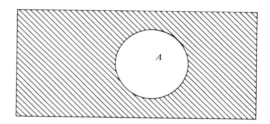

Finally, the *complement* of a set *A*, denoted by $\overline{A}$, is illustrated in Figure 5.4. The shaded area includes all elements of *U* that are not in *A*. Thus, again continuing our example with children, $\overline{A}$ would include all children that are not less than 10 years old. A particular child is a member of $\overline{A}$ to the extent that the condition of being less than 10 years old is *not* satisfied.

Fuzzy Sets

As we mentioned earlier, it often may be difficult to determine whether or not a particular element belongs to a set. Part of this difficulty is due to the fact that classical (i.e., nonfuzzy) set theory requires one to answer the question of membership with yes or no. Thus, for example, a classical perspective requires one to decide if the risk resulting from a particular site for a nuclear power plant is acceptable or unacceptable. It is quite easy to imagine someone wanting to answer such a question with a phrase such as "somewhat acceptable."

Fuzzy set theory was introduced by Lofti Zadeh in the 1960s (Zadeh 1965, 1973) with a goal of allowing one to represent situations where membership in sets cannot be defined on a yes/no basis. The *membership function* is a central concept in fuzzy set theory. Within this chapter, we shall use $\mu_A(x)$ to denote the membership of an element with property x in the fuzzy set *A*. Several examples are illustrated in Figure 5.5

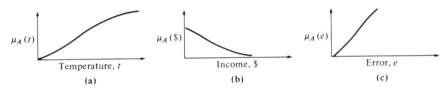

Figure 5.5. Membership functions for fuzzy sets of (a) hot temperatures, (b) poor people, and (c) unacceptable errors.

Membership functions usually are such that $0 \leqslant \mu_A(x) \leqslant 1$. A special case is where $\mu_A(x)$ can only have values of 0 or 1. In this situation, fuzzy set theory reduces to classical set theory. The value of $\mu_A(x)$ for a particular element with property x is referred to as the element's *grade of membership*. Alternatively, the value of $\mu_A(x)$ can be viewed as denoting the *degree of compatibility* between having property x and belonging to fuzzy set A.

If $\mu_A(x)$ takes on quantitative values, such as $0 \leqslant \mu_A(x) \leqslant 1$, then it is referred to as a type 1 membership function. On the other hand, if the values of $\mu_A(x)$ are words (e.g., "low") then we have a type 2 membership function. In this case the values of $\mu_A(x)$ (i.e., the words) would have to be defined by another membership function. Actually, a type n membership function is theoretically possible with somewhat of a hierarchical arrangement of levels of specificity. However, within this chapter, we shall only consider type 1 membership functions.

Before considering how basic set theory operations can be performed with fuzzy sets, it is useful to briefly compare the concepts of membership functions with that of subjective probability. Zadeh views membership functions as defining the *possibility* of an element belonging to a fuzzy set. From this perspective, membership may, for example, be possible but highly improbable.

Returning again to our example with children, consider a particular child who is 8 years old. Does this child belong to the fuzzy set of "young" children? With the membership function concept, we might say that it is reasonably possible that the child does belong to the set. However, there are no probabilities involved. We know the child's age. Further, the issue of the child belonging or not belonging to the set cannot be resolved by viewing membership as a random process such as playing roulette whereby some 8 year olds become members while others do not, or where a quirk of the random number generator may result in 9 year olds becoming members whereas 8 year olds do not become members. Thus, the notion of probability is not particularly useful in the example. However, possibility is a very useful concept.

Operations with Fuzzy Sets

While there are a wealth of operations that can be performed using fuzzy sets, including fuzzy statistics and fuzzy differential equations, we shall limit our discussions in this chapter to five basic operations. The first three of these are union, intersection, and complementation.

Let $\mu_A(x)$ and $\mu_B(x)$ denote the membership of an element with property x in fuzzy sets A and B, respectively. The union of fuzzy sets A and B, denoted by $A \cup B$, results in a fuzzy set such that an element with property x has membership

$$\mu_{A \cup B}(x) = \max[\mu_A(x); \mu_B(x)]. \tag{5.1}$$

Thus, an element with property x belongs to fuzzy set $A \cup B$ to the extent that it belongs to *either* fuzzy set A *or* fuzzy set B. In the special case that $\mu_A(x)$ and $\mu_B(x)$ can only have values of 0 or 1, we are dealing with nonfuzzy (i.e., classical) sets and (5.1) still holds. In this case, if either $\mu_A(x) = 1$ or $\mu_B(x) = 1$ or both, then $\mu_{A \cup B}(x) = 1$.

The intersection of fuzzy sets A and B, denoted by $A \cap B$, results in a fuzzy set such that an element with property x has membership

$$\mu_{A \cap B}(x) = \min[\mu_A(x); \mu_B(x)]. \tag{5.2}$$

Therefore, an element with property x belongs to fuzzy set $A \cap B$ to the extent that it belongs to *both* fuzzy set A *and* fuzzy set B. As before, (5.2) is is still valid for the special case of classical sets. In this situation, both $\mu_A(x)$ and $\mu_B(x)$ must equal 1 for $\mu_{A \cap B}(x)$ to equal 1.

The complement of fuzzy set A, denoted by $\overline{A}$, results in a fuzzy set such that an element with property x has membership

$$\mu_{\overline{A}}(x) = 1 - \mu_A(x). \tag{5.3}$$

For this operation, it is easy to see that the special case of classical sets results in (5.3) still being valid.

The fourth and fifth basic operations with fuzzy sets which we shall consider are *fuzzy relations* and *fuzzy compositions*. These operations are not quite as straightforward as union, intersection, and complementation. To explain the idea of fuzzy relations, consider two fuzzy sets A and B whose membership functions are shown in Figure 5.6. Using these membership functions, we should like to consider the relationship between the heaviness of an individual and the slowness of his running speed.

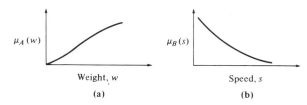

Figure 5.6. Membership functions for fuzzy sets of (a) weights that are heavy and (b) running speeds that are slow.

The membership of a particular weight w and a particular speed s in the relationship between fuzzy sets A and B is denoted by $\mu_{A \times B}(w, s)$ and given by

$$\mu_{A \times B}(w, s) = \min[\mu_A(w); \mu_B(s)]. \tag{5.4}$$

Using (5.4) and the membership functions shown in Figure 5.6, we see that very heavy weights and very slow speeds produce much higher values of $\mu_{A \times B}(w, s)$ than results with either nonheavy weights or nonslow speeds.

Another way of viewing $\mu_{A \times B}(w, s)$ is to consider it to be a *compatibility function*. Thus, the value of $\mu_{A \times B}(w, s)$ is the degree to which particular values of w and s are compatible, within the relationship between heavy weights and slow running speeds. Note, however, that low values of $\mu_A(w)$ and $\mu_B(s)$ produce low values of $\mu_{A \times B}(w, s)$. It would seem that nonheavy weights and nonslow speeds should be compatible.

We can consider this issue by emphasizing that we are interested in the relationship between heavy weights and slow running speeds. Starting with the assumption that a relationship existed, the membership functions in Figure 5.6 were constructed. The formulation was far from arbitrary. For example, had we chosen to define B as the fuzzy set of running speeds that are fast, the orientation of the membership function in Figure 5.6b would be reversed and would monotonically increase with speed. Thus, (5.4) would tell us that very heavy weights and very fast running speeds are highly compatible. Obviously, this is ridiculous and points out how one must be careful in formulating relationships. Thus, one cannot use (5.4) and any arbitrary $\mu_A(x)$ and $\mu_B(y)$ to "discover" relationships between fuzzy sets.

Fuzzy compositions deal with conditional relationships. For example, we might be interested in the fuzzy set of speeds given that weight is somewhat heavy. To consider this example, we shall start by defining C as the fuzzy set of somewhat heavy weights. An appropriate membership function is shown in Figure 5.7. Also, we shall define D as the fuzzy set of speeds given that weight is somewhat heavy.

Figure 5.7.
Membership function for
fuzzy set of weights that
are somewhat heavy.

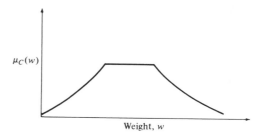

$\mu_C(w)$

Weight, w

To determine $\mu_D(s)$, we shall begin by considering a particular value of s and a particular value of w. Clearly, s can only be a member of fuzzy set D to the extent that w is a member of fuzzy set C and to the extent that s and w are compatible within the relationship between s and w. Thus, $\mu_D(s)$ for a particular w can be determined using $\min[\mu_C(w); \mu_{A \times B}(w, s)]$. Considering all w, s can be a member of fuzzy set D to the extent that *any* appropriate value of w is possible. Thus, a union operation is appropriate and we obtain

$$\mu_D(s) = \max\{\min[\mu_C(w); \mu_{A \times B}(w, s)]\}. \tag{5.5}$$

Applications

Fuzzy set theory has not been extensively applied to modeling human–machine interaction in realistic tasks. Thus, there is not a breadth of experience on which to base conclusions about the range of applicability of fuzzy set theory within human–machine systems. Nevertheless, the two applications which we shall discuss in this chapter do successfully illustrate the usefulness of fuzzy set theory. Further, they allow us to speculate about the characteristics of applications particularly amenable to fuzzy set formulations.

Basically, applications of interest involve situations where the human must cope with an inexact knowledge of the process being observed and/or controlled. Further, fuzzy set theory may be especially useful when the investigator who is studying the human–machine system does not have an exact knowledge of the process (i.e., the machine). In such situations, the investigator may have to rely on the human operator's verbal descriptions of the process and the rules for controlling it. The following application illustrates this type of situation.

A Process Control Model

Considerable research has been devoted to study of the process control operator (Edwards and Lees 1974). By process control, we mean human–machine systems such as chemical plants, refineries, and so on rather than automobiles, aircraft, ships, and other vehicle systems. Process control is typically more difficult to deal with because the equations which describe, for example, the chemical process of interest are not straightforward linear state equations such as those applicable to many vehicle systems. Thus, it is difficult to employ the estimation and control theory models discussed in Chapters 2 and 3.

However, even though it is difficult to describe the process of interest mathematically, the human operator is usually able to control it. Thus, one way to obtain a description of the process is to ask the human. The problem with this approach is that the human, at best, will only be able to provide a verbal, qualitative description of how the process is controlled. This type of description is not particularly useful within a linear state equation approach to modeling human–machine interaction. However, it quite nicely fits into a fuzzy set formulation. Tong (1977) has reviewed fuzzy set approaches to process control problems. Within this chapter, we shall consider the work of King and Mamdani (1977).

King and Mamdani studied single-input, single-output process control situations. The human operator was assumed to observe the error between the desired and actual process output, denoted by E, as well as the rate of change of the error, denoted by CE. Upon observing E and CE, the operator was assumed to respond with a change in control input CU. E, CE, and CU were assigned grades of membership in various fuzzy sets. For example, E was assigned membership grades in eight fuzzy sets: positive big (PB), positive medium (PM), positive small (PS), positive nil (P0), negative nil (N0), negative small (NS), negative medium (NM), and negative big (NB). The membership grades were quantified heuristically so that the performance resulting with the eventual fuzzy control algorithm was optimized.

Using analyses of verbal protocols as well as the fuzzy composition operation discussed earlier, King and Mamdani developed a control algorithm that includes various linguistic rules such as: if E is PB or PM, then if CE is NS, CU is NM. The composition operation was used to determine the overall membership for each possible control action. The control action with highest membership was then implemented.

The resulting control algorithm was applied to the control of a boiler and steam engine. It was compared to an algorithm based on conventional control theory. The final results, after each algorithm was optimized, was that the fuzzy algorithm yielded better system performance than the conventional algorithm. Thus, capturing the human's knowledge in a fuzzy set formulation led to performance improvements that would have other-wise been unlikely.

A Fault Diagnosis Model

As systems become increasingly complex and more automation is introduced, the human's role is shifting from that of an operator to that of a
monitor or supervisor (Sheridan and Johannsen 1976). One particularly
important aspect of the supervisory role involves dealing with the system
when it fails. Included here is failure detection, diagnosis, and correction as
well as operation of the system in a degraded mode. We discussed failure
detection in Chapter 2 and shall briefly return to that topic in Chapter 6.
Within this chapter, we shall consider failure diagnosis.

Rouse and his colleagues have pursued an extensive series of empirical
and theoretical studies of human problem solving performance in fault
diagnosis tasks. Two models of fault diagnosis performance were developed: one based on fuzzy set theory and the other employing a form of
production system model. At this point, we shall review the fuzzy set model.
The production system model will be considered in Chapter 6.

Based on a careful study of the important characteristics of fault
diagnosis tasks, a series of tasks were developed, including two context-free
tasks as well as several context-specific tasks. The use of context-free tasks
was motivated by a desire to study whether or not humans could develop
general, context-independent fault diagnosis abilities. Although this is a
fascinating issue (Rouse 1978a, 1979b,c; Hunt 1979), it is nevertheless
beyond the scope of this book. Our purpose here is to review the fuzzy set
model of human decision making in the two context-free fault diagnosis
tasks (Rouse 1978b, 1979a).

An example of task 1 is shown in Figure 5.8. This display was generated
on a Tektronix 4010 by a DEC System-10. These networks operated as
follows. Each node or component had a random number of inputs.
Similarly, a random number of outputs emanated from each component.
Components were devices that produced either a 1 or 0. Outputs emanating
from a component carried the value produced by that component. A
component would produce a 1 if *all* inputs to the component carried values
of 1, and the component had not failed. If either of these two conditions
were not satisfied, the component would produce a 0. Thus, components
were like AND gates. If a component failed, it would produce values of 0 on
all the outputs emanating from it. Any components that were reached by
these outputs would in turn produce values of 0. This process continued
and the effects of a failure were thereby propagated throughout the
network.

A problem began with the display of a network with the outputs
indicated, as shown on the right-hand side of Figure 5.8. Based on this
evidence, the subject's task was to "test" connections until the failed

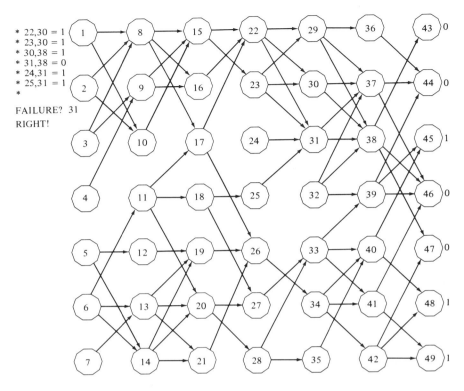

Figure 5.8. An example of task 1.
Based on Rouse (1978a).

component was found. All components were equally likely to fail, but only one could fail within any particular problem. Subjects were instructed to find the failure in the least amount of time possible, while avoiding all mistakes and not making an excessive number of tests.

The upper left-hand side of Figure 5.8 illustrates the manner in which connections were tested. An asterisk was displayed to indicate that subjects could choose a connection to test. They entered commands of the form k_1, k_2 and were then shown the value carried by the connection. If they responded with a simple "return," they were asked to designate the failed component. Then they were given feedback about the correctness of their choice. After this the next problem was displayed.

Task 1 was fairly limited in that only one type of node or component was considered. Further, all connections were feed-forward and thus, there were no feedback loops. To overcome these limitations, a second trouble-shooting task was devised.

Figure 5.9 illustrates an example of task 2. As with task 1, inputs and outputs of components could only have values of 1 and 0. A value of 1 represented an acceptable output whereas a value of 0 represented an unacceptable output.

A square component would produce a 1 if *all* inputs to the component carried values of 1, and the component had not failed. If either of these two conditions were not satisfied, the component would produce a 0. Thus, square components were like AND gates.

A hexagonal component would produce a 1 if *any* input to the component carried a value of 1, and the component had not failed. As before, if either of these two conditions were not satisfied, the component would produce a 0. Thus, hexagonal components were like OR gates.

As with task 1, all components were equally likely to fail, but only one component could fail within any particular problem. Subjects obtained information by testing connections between components (see upper left of Figure 5.9). Tests were of the form k_1, k_2, where the connection of interest was an output of component of k_1 and an input of component k_2. The

Figure 5.9.   An example of task 2.
Based on Rouse (1979a).

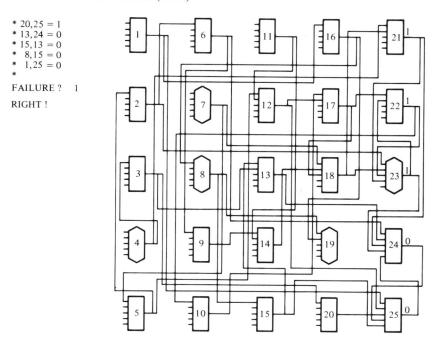

instructions to the subjects were the same as used for task 1. Namely, they were to find the failure as quickly as possible, avoid all mistakes, and avoid making an excessive number of tests.

Now, let us consider a model of problem solving in the above two fault diagnosis tasks. The task of fault diagnosis involves partitioning the set of all components into the feasible subset (i.e., those components that could possibly be the source of the unacceptable outputs) and the infeasible subset (i.e., those components that could not possibly be the source of the unacceptable outputs). From a purely technical point of view, all components belong to either the feasible subset or the infeasible subset. However, from a behavioral point of view, it is very difficult for the human to make such a strict partitioning. Fuzzy set theory offers a quite useful approach to representing the human's performance in this partitioning task.

When a task begins, one has knowledge of several outputs. The 0 outputs are the symptoms of the failure. Assuming that only one component has failed, the members of the feasible subset are those components that could cause *all* of the symptoms. Using a fuzzy set approach to defining this subset, let the membership of component i in the fuzzy subset of feasible causes of the 0 output of component j be given by

$$\mu_{ij}^0(x_{ij}) = 1/(1 + \alpha x_{ij}^2), \qquad (5.6)$$

where x_{ij} is the "psychological distance" between components i and j and α is a free parameter.

If there are no feedback loops, (5.6) is sufficient to define the membership of component i in the fuzzy subset of possible causes of all known 0 outputs. However, the possibility of feedback presents difficulties because a 0 output at component i does not necessarily mean that component i is the source of the symptom at component j. It is possible that the symptom is being fed back from j to i. To consider this possibility, define the membership of component i in a feedback loop from component j as

$$\mu_{ji}^f(x_{ji}) = 1/(1 + \gamma x_{ji}^2), \qquad (5.7)$$

where x_{ji} is as defined earlier and γ is a free parameter.

To form the fuzzy subset of feasible solutions, one is interested in components that can cause *all* known 0 outputs while also *not* being involved in feedback loops from the 0 outputs. In other words, although one is interested in components that can cause all known 0 outputs, one also wants to avoid valueless tests that simply measure these 0 outputs as they are fed back into the network. This feasible subset can be defined in three steps. First, one forms the intersection of the fuzzy subsets defined by (5.6). Then, one forms the intersection of the complements of the fuzzy subsets defined by (5.7). Finally, the feasible subset is formed by intersecting the two resulting fuzzy subsets. This leads to an overall membership of component i in the fuzzy subset of feasible solutions given by

$$\mu_i^0 = \min_j [\mu_{ij}^0; 1 - \mu_{ji}^f]. \tag{5.8}$$

While the symptoms (the 0 outputs) have now been considered, the 1 outputs also provide very useful information. The membership of component i in the fuzzy subset of contributors to the 1 output of component j will be defined as

$$\mu_{ij}^1(x_{ij}) = 1/(1 + \beta x_{ij}^2), \tag{5.9}$$

where x_{ij} is defined as above and β is a free parameter.

Since feedback of 1 outputs is only an important consideration if the network is dominated by redundancy (i.e., OR components), feedback from j to i can be ignored in the case of 1 outputs. One is only interested in the possibility of component i contributing to *any* of the 1 outputs. Thus, the union of fuzzy subsets defined by (5.9) is desired. This leads to an overall membership of component i in the fuzzy subset of infeasible solutions given by

$$\mu_i^1 = \max_j [\mu_{ij}^1]. \tag{5.10}$$

Considering the overall fault diagnosis problem, one is interested in components with strong membership in the feasible subset and weak membership in the infeasible subset. The fuzzy subset of such components can be defined by forming the intersection of the feasible set with the complement of the infeasible set. The membership of component i in the resulting fuzzy subset is given by

$$\mu_i = \min[\mu_i^0; 1 - \mu_i^1]. \tag{5.11}$$

Knowing the above membership function for all components, one can choose to test the component whose membership function is maximum.

In order for the fuzzy algorithm specified by (5.6)–(5.11) to realistically represent human behavior, the algorithm had to be supplemented with various nonfuzzy heuristics. For task 1, heuristics included rules for eliminating components from further consideration as well as a stopping rule that enabled the algorithm to know when it was finished. For task 2, a heuristic was needed to handle OR components. This heuristic took the form of modifying coefficients on the membership functions specified by (5.6), (5.7), and (5.9). A stopping rule was also needed for task 2.

The final ingredient in this model is the measure of psychological distance. After considering a variety of issues (Rouse 1978b), the following measure was proposed. For a given 0 or 1 output, all components that are functionally related to the output of interest are ranked in terms of shortest geographical distance along the path connecting each component to the output. The psychological distance of each component from the output of interest is then defined as the rank which the component has in the ordering. As components are tested and the values of their outputs become known, they are removed from the rank ordering. In this way, the psychological distances of the remaining components decrease and hence their grades of membership increase.

In three experiments the performance of the fuzzy set model was compared with human performance in terms of number of tests until problem solution. In the first experiment, eight subjects performed a self-paced version of task 1. A forced-pace version of task 1 was employed for the second experiment, which included 12 subjects. The third experiment included four subjects performing task 2. The model's parameters (i.e., α, β, and γ) were adjusted in order to match model performance with that of the subjects. For the task 1 data, the model was matched to the average across subjects while, for the task 2 data, the model was matched to each subject.

To interpret the resulting values of α, β, and γ, one must first note that large values of these parameters cause the membership functions to decrease more rapidly, as a function of psychological distance, than with small values. Thus, as the values of the parameters decrease, the model assumes that the number of relationships within the network that enter into the human's solution strategy increases. Therefore, performance should improve as α, β, and/or γ decrease.

Considering all three sets of data, two general conclusions were reached. First, subjects were reasonably good at finding components that were feasible in terms of reaching all of the symptoms (i.e., the 0 outputs). Second, subjects were fairly poor in terms of considering how the 1 outputs should impact their search. In other words, β tended to be relatively large. To test this notion further, subjects were provided with a computer aid that took account of the 1 outputs for them by iteratively crossing off components (i.e., by drawing Xs through them) that reached known 1 outputs and, for that reason, were not feasible solutions. As one might expect, subjects' performance improved substantially. In addition, the fuzzy set model was able to predict this improvement without any adjustment of its parameters. Thus, the model has potential for assessing the benefits of various computer aids.

The data for task 2 indicated substantial individual differences among subjects. Fitting the model to each subject, it appeared that the main difference among subjects was the way in which they considered feedback loops. Subjects who tended to discount the importance of feedback loops (i.e., relatively large γ) were found to have difficulty on problems with many feedback loops. Thus, it appears that the model is a useful tool for inferring individual strategies and, in that way, may prove of value in designing computer aids that are customized to particular individuals.

To conclude this discussion of fault diagnosis, it is quite useful to note how the fuzzy set model allowed interpretation of human behavior in two rather complex tasks to be reduced to a problem of interpreting the resulting values of the three parameters of the model. As we discussed in Chapter 1, this use of a model to provide a succinct description of a complex phenomenon is one of the main benefits of the modeling process. Further, this benefit has ramifications beyond mere description. For example, the behavioral interpretations possible with the fuzzy set model of fault diagnosis had a direct impact on the design of a computer-based training system which has been implemented within an aircraft power plant maintenance curriculum at the University of Illinois (Hunt 1979; Johnson 1979). Finally, we hasten to note that success in using models to provide succinct behavioral descriptions often depends upon the somewhat serendipitous discovery of useful intervening variables such as, for example, psychological distance.

Summary

Within this chapter, we have viewed the human as a logical problem solver. Fuzzy set theory has been proposed as a means of representing the human in that role. Most importantly, the concept of membership functions was suggested as a means of representing constraints upon the human's abilities as a problem solver. This notion led us to describe human problem solving in terms of fuzzy sets that are manipulated using nonfuzzy operations. Although it is, of course, possible that the logical operations employed by the human are also fuzzy, this extra degree of freedom does not appear to be necessary to provide adequate representations of human behavior, at least in the applications considered in this chapter.

Besides discussing the concept of membership functions, we also considered five basic fuzzy set operations: union, intersection, complementation, relation, and composition. As we noted earlier, these basic operations are only a small portion of what is available in the immense literature on fuzzy sets. Nevertheless, these basic operations have proven quite sufficient for the applications discussed in this chapter, namely, process control and fault diagnosis.

Fuzzy set theory is quite attractive because it allows us to consider situations that are describable in only qualitative, verbal terms. Further, a rigorous calculus is provided with which one can manipulate the resulting fuzzy sets. However, we by all means do not want to leave the reader with the impression that fuzzy set formulations are always straightforward. For example, from a behavioral perspective, it can be quite difficult to define the independent variable or variables of the membership functions. Thus, although fuzzy set theory allows one to manipulate membership in various sets as a function of x, one of the behavioral scientist's main problems may be defining and measuring x.

Chapter 6
Production Systems, Pattern Recognition, and Markov Chains

The theories of estimation, control, queueing, and fuzzy sets are well developed in the sense of being based on comprehensive collections of assumptions, derivations, and theorems. Further, the notion of optimization is highly developed in these theories. Finally, and perhaps most important from a practical point of view, these theories are reasonably straightforward to use because the computational procedures necessary to their implementation have been developed. These three reasons have led to the use of estimation, control, queueing, and/or fuzzy sets in a variety of human–machine systems domains and, in that way, justified devoting a chapter in this book to each of these theories.

In this chapter, we shall consider three modeling methodologies that have not, as yet, received extensive use in system engineering as applied to human–machine interaction. However, these methods have been used extensively in other domains, either in the sense of having been employed to develop behavioral models or having been used in purely engineering applications. Production systems and Markov chains are an example of the former situation whereas pattern recognition is an example of the latter case. Nevertheless, these methods are appropriate for modeling human–machine interaction and deserve consideration in this book.

Production Systems

The field of artificial intelligence has provided an abundance of interesting and sometimes controversial ideas (Newell and Simon 1972; Winston 1977; Weizenbaum 1976). From an engineering perspective, one of the most useful notions is that of production systems as advanced by Newell. Basically, a production[1] is a situation–action pair where the situation side is a list of things to watch for and the action side is a list of things to do. A production system is a rank-ordered set of productions where the actions resulting from one production can result in situations that will cause other productions to execute. In other words, a production system is a rank-ordered set of pattern-evoked rules of action such that the actions modify the pattern and thereby evoke other actions. Production system models are often also called rule-based models.

To illustrate the use of production system models, we shall consider several examples, starting with Newell's model of human performance in reaction time tasks (Newell 1973). Newell's view of human information processing is depicted in Figure 6.1. Long-term memory (LTM) is seen as composed entirely of an ordered set of productions whereas short-term memory (STM) holds an ordered set of symbolic expressions. The model processes information by observing the contents of the STM on a last

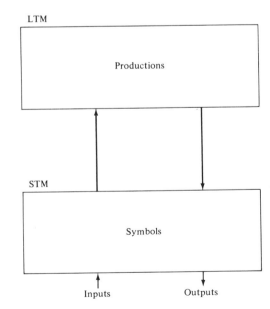

Figure 6.1.
Newell's information processing model.
Based on Newell (1973).

---

[1] We hasten to note that the word "production," as it is used here, has absolutely nothing to do with the manufacturing connotation of the word.

come–first served basis. A match occurs when a symbol or symbols in the STM match the situation side of a production in the LTM. Then an action is evoked which results in new symbols being deposited in the STM. This process of pattern-evoked actions goes on continually and, as a result, people play chess, solve arithmetic problems, and so on (Newell and Simon, 1972).

It is interesting to contrast this information processing analogy of how the human manipulates symbols with the use of information theory to describe human performance. Based on Shannon and Weaver's classic work (1948), a variety of investigators have employed information theory to describe how the human employs a priori probabilities and subsequent "messages" to cope with uncertainty (e.g., Attneave 1959; Sheridan and Ferrell 1974, Chaps. 5–7). For somewhat simplified situations involving, for example, reaction time and absolute judgment tasks, information theory metrics have worked quite well. However, information theory has mainly served as a measure rather than as a model.

Newell's information processing model is quite different in that it describes the process of transforming information (i.e., symbols) to decisions and actions. Further, it is not particularly concerned with uncertainty. In other words, production system models are process theories whereas information theory models are statistical theories. For the above reasons, it should be clear that information processing as discussed here is only slightly related to information theory in the usual sense.

Newell (1973) employed his information processing model to explain the classical results obtained with a task developed by Sternberg. The task is such that subjects memorize a small set of symbols (e.g., digits). They are then presented with a particular symbol and asked to respond positively if this symbol is in the memorized list and negatively otherwise. The classical result is that reaction times for positive and negative responses are similar. This result would lead one to believe that subjects exhaustively search the memorized list regardless of the location of the displayed symbol in the list. Otherwise, if the search stopped when a subject found a match, positive responses would require only half the time of negative responses since, for the average positive response, subjects would find a match halfway through the list.

Using a production system approach, Newell studied a wide variety of hypotheses that might explain this apparent exhaustive search. He eventually settled on an encoding/decoding hypothesis. Succinctly, the idea is that the goal of reliability motivates the information processing system to encode the memorized list into a single chunk. Thus, when a displayed symbol is presented, the entire list must be decoded. Therefore, the subject appears to be searching exhaustively but is actually decoding. In contrasting this production system model with available reaction time data for the

Sternberg task, a favorable comparison resulted. To further validate Newell's model, error data would also have to be compared. For success to result from such a comparison, the model would have to be revised so that it would in fact make errors.

Although Newell's results are certainly interesting from a behavioral perspective, the level at which he has approached the problem presents difficulties from a systems engineering point of view. The main difficulty is that modeling, for example, piloting of a realistically complex aircraft in terms of transformations of the contents of the pilot's STM, would be an overwhelming task. We need a more macroscopic approach. Fortunately, the production system concept is still useful at such a level.

Wesson (1977) has proposed a production system model for the task of air traffic control. Some example productions are shown in Figure 6.2. One particularly notable aspect of this model is the fact that its rank ordering of productions is situation dependent. Thus, for example, emergencies may result in different priorities than used in normal situations. The computer program in which Wesson's model was embodied was found to perform in a manner comparable to real air traffic controllers.

Goldstein and Grimson (1977) have suggested a production system model for skill acquisition with application to attitude instrument flying. Their model incorporates approximately 50 productions. The productions for achieving level flight are shown in Figure 6.3. Besides the usual production system structure, Goldstein and Grimson's model also includes what they call annotations. This allowed them to incorporate, within their program, rationales and caveats for particular productions. Their model is especially of note because of their efforts to incorporate learning within the production system approach.

In Chapter 5, we discussed a fuzzy set model of human performance in fault diagnosis tasks. This model was successful in the sense that it

Figure 6.2.
Example productions in Wesson's
air traffic control model.
Based on Wesson (1977).

1. Separation conflict ⟶ try climbing
                                           try descending
                                           try turning
                                           try slowing down
                                           try holding

2. Too high for approach handoff ⟶ descend

3. Off track ⟶ try turning back to track
                         ˌ try next radio fix
                         try doing nothing

1. Achieve level flight
2. Notice △ pitch via artificial horizon
3. Notice △ pitch via vertical velocity indicator
4. Notice △ pitch via altimeter
5. Notice △ pitch via airspeed
6. Eliminate △ pitch with elevators
7. Eliminate △ pitch with throttle

Figure 6.3.

Goldstein and Grimson's productions for level flight.

Based on Goldstein and
Grimson (1977).

mimicked subjects in terms of making the same number of tests to solve a problem. Based on this success, the next natural question was the following: Does it make the *same* tests as subjects? To pursue this issue, the test sequences of various subjects were studied. From this study, the idea of using a production system model emerged. The model is shown in Figure 6.4. This model represents the process of extracting patterns of problem structure and symptoms of failures, and evoking of rules of thumb for choosing tests. For example, rule 1 in Figure 6.4 is a stopping rule that looks at the state of the network and asks, "Have I found the failure?"

This model was used to describe the performance of subjects in the two fault diagnosis tasks discussed in Chapter 5. (See Figure 5.8 and 5.9.) For task 1, it was found that a rank-ordered set of 13 rules could describe almost all of the tests made by subjects (Pellegrino 1979). Example rules include:

Find a component with a 0 output and test its inputs,

Find a component common to all symptoms and test its inputs,

Test inputs to the component whose output was just found to be zero.

Extensions of this work to task 2 found that a much smaller, and more general, set of rules could adequately describe test choices in task 2. The smaller set was somewhat more useful in that it was easier to interpret various patterns of behavior. For both tasks, it was found that the rank ordering of rules could be used to explain how different methods of training allowed particular subjects to perform better than other subjects (Rouse et al. 1980).

The four production system models presented here can serve to illustrate an important feature of many models in the field of artificial intelligence. Notice that none of these models emphasized statistical measures of performance. Instead, they tried to emulate sequences of actions. As Gregg and Simon (1967) point out, this is an ambitious approach, fraught with many more difficulties than the statistical approach. Nevertheless, modeling sequences of actions rather than statistics of performance can potentially

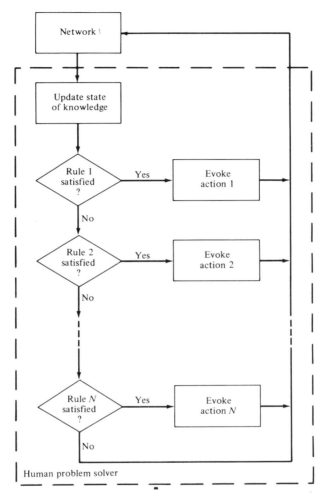

Figure 6.4. A production system model of fault diagnosis.
Based on Rouse et al. (1980).

provide considerable insight into human behavior. From an engineering perspective, this insight may be particularly useful in terms of providing better structures for statistical models.

Pattern Recognition

Production system models view human behavior in terms of how situations or patterns of events evoke actions. It is interesting to consider how the human recognizes patterns. The literature of pattern recognition and artificial intelligence addresses this question. Recent reviews in the pattern recognition domain include IEEE (1977), and Sklansky (1978), whereas Winston (1975, 1977) considers the artificial intelligence area.

Two approaches to pattern recognition have received particular attention: statistical methods and syntactical methods. The statistical methods use discriminant functions to classify patterns. This involves extracting a set of features from the pattern and statistically determining how close this feature set is to the a priori known features of candidate classes of patterns. The class whose features most closely match the measured features is chosen as the match to the pattern of interest, with of course some consideration given to the a priori probabilities of each class and the costs of errors.

The syntactic methods partition each pattern into subpatterns or pattern primitives. It is assumed that a known set of rules (a grammar) is used to compose primitives into a pattern. One approach to recognizing primitives is to use the statistical approach noted above.

Another aspect of pattern recognition involves image processing. Here, each picture element (pixel) is classified according to gray level. Then thresholds are used to segment the picture. More elaborate approaches use multidimensional classification of each pixel and then use an appropriate multidimensional clustering of similar pixels.

Artificial intelligence researchers have devoted considerable effort to scene analysis. With emphasis on understanding scenes composed of somewhat arbitrary collections of blocks, methods have been developed to pick particular blocks out of scenes, even if the desired block is partially hidden.

Most of the methods discussed above have worked reasonably well within limited domains. When the context within which one is working is well understood, it is often possible to successfully sense and interpret inputs, although considerable computational power may be needed.

Although the advent of inexpensive microelectronics might allow one to utilize large amounts of computational power in a model of human pattern recognition, there are bigger problems to be solved. Namely, it is difficult to deal with realistic contexts in a static manner. What one sees depends on what one looks for, what one expects to see, and the costs of not seeing it. These aspects of seeing cannot be considered out of context and without reference to the specific individual involved. From a theoretical point of view, one might represent context in terms of a priori probabilities with perhaps some form of Bayesian updating (Curry 1971). However, although this approach is theoretically tenable, it does present enormous measurement problems.

The above discussion has served to emphasize the point that developing an operationally useful model of the human as a general pattern recognizer is a formidable task. Nevertheless, pattern recognition methods can be employed to achieve more modest modeling goals. Within this chapter, we shall consider discriminant analysis, one of the more elementary pattern recognition methods. Thorough treatments of discriminant analysis can be found in Tatsuoka (1971) as well as Afifi and Azen (1972).

We shall explain discriminant analysis in the context of display monitoring. Suppose that a human is monitoring a display, perhaps looking for the occurrence of some event. For example, an air traffic controller might be looking for deviations of aircraft from their commanded flight paths. It seems reasonable to assume that the human extracts various features, x_j, $j = 1, 2, \ldots, p$, from observations. These features are properties of the observations that the human feels characterize the presence or absence of events. For example, the air traffic controller might extract aircraft speed, rate of turn, and so on.

Following the extraction of a set of features, the value of a linear discriminant function

$$Y = v_1 x_1 + \cdots + v_p x_p \tag{6.1}$$

is calculated. It is assumed that the human's experience with the task makes it possible to develop coefficients v_j, $j = 1, 2, \ldots, p$, that result in the discriminant score Y that best differentiates observations of events from observations of nonevents.

This model is attractively simple. The critical issue in modeling human performance in various situations of interest is the choice of features to be employed. Although, as we considered within our discussion of process control in Chapter 5, this issue can partially be resolved by asking the human for a verbal explanation of how he performs the task. Thus discriminant analysis is similar to fuzzy set theory, and different than the theories of estimation, control, and queueing, in that it allows straightforward incorporation of the human's qualitative statements.

With the features chosen, the next issue is determination of v_j, $j = 1, 2, \ldots, p$. Following the derivation of Tatsuoka (1971), let x_{ijk} be the value of the jth feature extracted from the ith set of observations when an event was not ($k = 1$) or was ($k = 2$) present, where $i = 1, 2, \ldots, n_k$. Let $X^{(k)}$ denote the n_k by p matrix of observations for group k.

The centroid of the kth group is a p vector, denoted by $\overline{X}^{(k)}$, whose elements are formed by averaging across the n_k observations of each feature. Thus, the jth element of $\overline{X}^{(k)}$, denoted by $\overline{x}_j^{(k)}$, is given by

$$\overline{x}_j^{(k)} = \frac{1}{n_k} \sum_{i=1}^{n_k} x_{ijk}. \tag{6.2}$$

The sum of squares and cross-products matrix of the kth group is a p by

p matrix, denoted by S_k, with elements $(s_{ij})_k$ given by

$$(s_{ij})_k = \sum_{l=1}^{n_k} [x_{lik} - \overline{x}_i^{(k)}][x_{ljk} - \overline{x}_j^{(k)}]. \tag{6.3}$$

The covariance of the kth group is then given by

$$D_k = S_k/(n_k - 1). \tag{6.4}$$

The total sums of squares and cross-products matrix is a p by p matrix, denoted by $\mathbf{T}$, with elements t_{ij} given by

$$t_{ij} = \sum_{k=1}^{2} \sum_{l=1}^{n_k} [x_{lik} - \overline{x}_i][x_{ljk} - \overline{x}_j], \tag{6.5}$$

where

$$\overline{x}_j = \frac{1}{n_1 + n_2} \sum_{k=1}^{2} \sum_{l=1}^{n_k} x_{ljk}. \tag{6.6}$$

We should like to determine the $V = (v_1, v_2, \ldots, v_p)$ that maximizes the differences between the average values of Y resulting with each of the two groups. It can be shown (Tatsuoka 1971) that this condition is satisfied if

$$V = T^{-1}(\overline{X}^{(1)} - \overline{X}^{(2)}) \tag{6.7}$$

which leads to group means of

$$\overline{Y}^{(k)} = \overline{X}^{(k)'} V \tag{6.8}$$

and group variances of

$$D_k(Y) = V'D_k V. \tag{6.9}$$

Given that V, D_k, and so on are known, we are now in a position to deal with a new set of observations. Suppose a set of features is extracted and, using (6.1), a value of Y results. Which group should we conclude that this observation came from? The a posteriori probability of group membership is given by

$$p(k|Y) = e^{-(\chi_k^2)/2} \bigg/ \sum_{l=1}^{2} e^{-(\chi_l^2)/2}, \tag{6.10}$$

where

$$\chi_k^2 = \frac{(Y - \overline{Y}^{(k)})^2}{D_k(Y)} + LN[D_k(Y)] - 2LN(p_k), \tag{6.11}$$

where p_k is the a priori probability of membership in group k. The values of $p(k|Y)$ can be combined with the costs of incorrect classification and the rewards of correct classification to yield a decision that maximizes expected reward or, equivalently, minimizes expected cost.

Greenstein and Rouse have employed this model for describing human event detection in multiple process monitoring situations (Rouse and Greenstein 1976; Greenstein and Rouse 1978; Greenstein 1979). The task which they considered is illustrated in Figures 6.5 and 6.6. Subjects simultaneously viewed the sampled outputs of nine second-order dynamic processes. Their task was to detect changes in the signal-to-noise ratio characterized by the process outputs becoming increasingly noisy. Their instructions were to detect changes as quickly as possible while also avoiding false alarms.

Considering Figure 6.5, events (i.e., the onset of the decreasing signal-to-noise ratio) had occurred in processes 1, 2, 3, and 8 at times 130, 113, 92, and 106, respectively. At the point at which this display was generated, the

Figure 6.5. The multiple process monitoring situation.

Based on Greenstein and Rouse (1978).

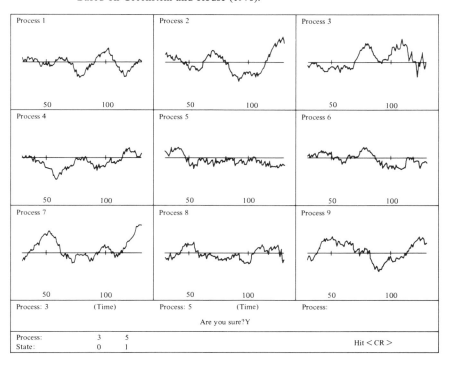

subject had only detected the event in process 3 while also incorrectly responding to process 5. Figure 6.6 illustrates the display 10 time units later (one update). The dashed vertical lines indicated to the subject the last point of response to each process.

In studying this event detection task, Greenstein and Rouse were primarily interested in developing a model for situations where knowledge of the dynamics of the processes was unavailable and thus one had to observe the human to determine how the task could be performed. Considering this constraint, the estimation theory approach of Gai and Curry presented in Chapter 2 was not applicable because their model assumed a knowledge of the process dynamics. On the other hand, the pattern recognition approach presented here was quite suitable.

To test the usefulness of the pattern recognition model, an experiment was conducted in which eight subjects performed the above event detection task for three trials, each of which was approximately 1 hour in duration.

Figure 6.6. An updated display.

Based on Greenstein and Rouse (1978).

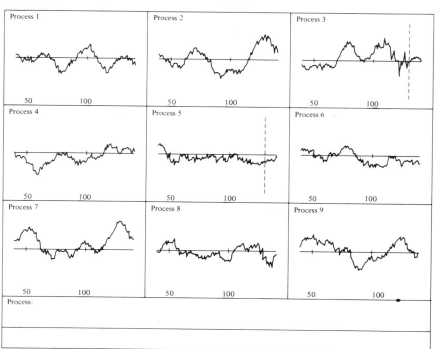

As they performed the task, subjects kept written "logs" of their actions and descriptions of what they were doing. Based on these logs, four features were selected for use in the model.

The first feature involved the magnitude changes in the sequence of recent process outputs. The second feature entailed the presence of reversals in direction in this sequence (i.e., changes of slope from positive to negative or vice versa). The third feature was based on the simultaneous occurrence of large-magnitude changes and reversals. Finally, the fourth feature was a very local measurement of magnitude changes in the most recent four points in the sequence of process outputs. In calculating average changes in magnitude, average number of reversals, and so on, an exponential averaging method was used to accommodate the fact that events became more pronounced as they evolved.

The model was tested by allowing it (i.e., the computer) to "watch" each subject during one of his or her trials. From these observations, the discriminant coefficient vector V was formed as well as the covariance matrix D_k, etc. The model's performance was then compared to that of the subject on another trial (i.e., one on which the model had *not* been trained). It was found that the model did quite well in terms of matching a subject's average time to event detection and number of correct detections, although it was somewhat more conservative than subjects in terms of false alarms.

These results have considerable practical implications, especially in the realm of computer-aided decision making. Consider a situation in which a human must simultaneously monitor many processes. Further, assume that a lack of knowledge of the dynamics of the processes (e.g., a chemical plant) as well as the presence of time-varying priorities and costs preclude direct automation of the monitoring tasks. Using the event detection model described above, a computer system could be designed to "watch" the human operator and hence be capable of providing back-up decision making if the human were to become overloaded. The decision of when the computer should help might be handled using the queueing theory formulation discussed in Chapter 4.

Considering further the use of pattern recognition methods within the domain of human–computer interaction, in Chapter 3 we discussed the problem of determining how a human has allocated attention among tasks. At that point, we discussed Enstrom and Rouse's work on applying fading-memory parameter estimation schemes to the problem of detecting changes of the human's input–output relationship in the control task. The algorithm presented in Chapter 3 was quite satisfactory for identifying the time-varying parameters of the human's input–output relationship. However, a method was needed for classifying the resulting parameters into, in that particular application, two classes of attention allocation: controlling only and controlling plus mental arithmetic. The above discriminant analysis model was applied to this problem and worked quite nicely, detecting in real time approximately 95% of the shifts of attention to the mental arithmetic tasks with few false alarms.

Another application of discriminant analysis to human-computer interaction is that of Weisbrod et al. (1977). This research was concerned with information gathering tasks. The discriminant function was viewed as a utility function with the coefficients reflecting varying priorities for different types of information. Once formed, the discriminant function was used to provide advice to the human in terms of how to select information so as to be consistent with the preferences identified by the pattern recognition model. It was found that the use of this aiding system considerably increased the number of decisions per unit time that a human could make.

As a final comment on pattern recognition, it is interesting to contrast the algorithm presented in this chapter with the parameter estimation algorithms considered in Chapter 2. The discussion in this chapter has mainly been concerned with how the human recognizes patterns associated with some class of events whereas Chapter 2 was primarily concerned with the estimation of the pattern of dynamic relationships among elements of time series. Although these two problems are somewhat different, the underlying mathematics are really fairly similar. Thus, pattern recognition and parameter estimation are by no means strictly different methods.

Markov Chains

Within this chapter, we first discussed production systems (i.e., rank-ordered sets of pattern-evoked rules of thumb) and then concerned ourselves with the problem of how patterns are detected. Thus, at this point, we have a notion both of how patterns are recognized and how they are utilized via production systems. Now, we want to consider the problem of modeling sequences of pattern-evoked actions, especially when these sequences are probabilistic in nature. This brings us to the topic of Markov chains, a term that we discussed briefly in Chapter 4 and now shall consider in more depth. Much of our discussion is based on Anderson and Goodman (1957), Howard (1960), and Ross (1970).

We shall first consider a process whose state $x(t)$ may take any of a discrete number of values denoted by 1, 2, ..., M. From a mathematical point of view, we need not be particularly concerned with the actual M possibilities for $x(t)$. As will later be illustrated, the states need not even be quantitative. Thus, $x(t)$ need only provide a nominal scale for the state of the process. Now suppose that the process can be observed at times $t = 0, 1, 2, \ldots$. Therefore, we have a discrete-time process.

We are interested in determining the probability that the state of the process equals j at time $t + 1$, denoted by $p[x(t + 1) = j]$. In general, determining this probability depends on knowing the state of the process at time t, $t - 1$, $t - 2$, and so on. However, if the state of the process at time $t + 1$ only depends on the state of the process at time t, then we have a first-order Markov process and the probability of interest is denoted by $p_{ij} = p[x(t + 1) = j \mid x(t) = i]$. The $M \times M$ matrix of transition probabilities P is called the transition matrix, which may be time varying. The transition matrix, along with the initial conditions, completely describe the first-order, discrete-time, finite Markov chain. This is one of the simplest Markov chain models.

This simple model can be elaborated in many ways. For example, a second-order Markov process would have transition probabilities of the form $p_{ijk} = p[x(t + 1) = k \mid x(t) = j, x(t - 1) = i]$. This model can be reduced to a first order if we redefine the state of the process as a composite of the state at time t as well as the state at time $t - 1$. In this way, higher-order models can be reduced to first-order models and thereby allow one to employ solution procedures developed for first-order processes. On the other hand, such reductions can present computational problems because forming composite states can cause M to become quite large. Nevertheless, higher-order Markov models can be quite useful for describing paths through the state space.

A continuous-time Markov process is such that state transitions can occur at any point in time. Such a process is described by transition rates rather than transition probabilities. Equation (4.1) illustrates the use of transition rates and, further, points out how queueing theory is based, to a large extent, on Markov models.

We can also consider Markov processes where the state is continuous rather than discrete. Such a process model is useful in describing stochastic linear dynamic systems (Meditch 1969). Thus, we find that much of the results in Chapters 2 and 3 were influenced by underlying Markov process models. The above discussions serve to emphasize the considerable dependence of the modeling methodologies described in this book on Markov process theory.

Developing a Markov chain model involves defining the states of the process, estimating the transition probabilities (or rates), and then perhaps making inferences about the "true" transition probabilities and the order of the process. Further, one might like to compare two Markov models and determine if they are significantly different. Anderson and Goodman (1957) developed a set of methods for considering such statistical questions for discrete-time finite Markov chains, and Billingsley (1961) developed methods for continuous-time models. The methods of Anderson and Goodman are quite straightforward and we shall consider their use later in our discussion.

Markov chain models have been quite popular with mathematical psychologists, as evidenced by the two-volume *Contemporary Developments in Mathematical Psychology*, edited by Krantz et al. (1974), which is heavily laced with Markov chains. For example, Greeno (1974) considers Markov chain models of learning. He discusses models that view learning as a transition through various stages. Greeno also considers Markov models of such classical conditioning situations as training an animal to avoid a shock.

Another example of the work of mathematical psychologists is that of Estes and Suppes (1974). These authors developed a formalization of stimulus sampling theory, an area of study within psychology that deals with how humans select from a presented set of elements in an effort to match or satisfy some relationship among elements that they are attempting to learn. Estes and Suppes present a very general formulation of stimulus sampling and show that it possesses desirable Markov chain properties. The general formulation is such that it can be specialized to any of a variety of types of learning situation.

On a more applied level, Penniman (1975) has developed a Markov chain model of human behavior when using on-line computerized bibliographic information retrieval systems. His model includes four states:

1. index search
2. logic formation
3. document display
4. other

He collected data and formed transition matrices as a function of four experimental variables, including length of terminal session, portion of session, data base accessed, and particular users. He then compared the resulting transition matrices using the statistical methods of Anderson and Goodman. He found that all four experimental variables significantly affected the transition probabilities, although he had to compare inexperienced to experienced users in order to obtain significant differences among users. Also, he found that higher-order models provided significantly better levels of predictability than lower-order models. Thus, the path by which a user arrived at a particular state was important.

Hammer and Rouse (1979) have studied how humans use on-line editors in preparing computer programs and text on a time-shared computer. They were interested in how different editing systems affected users' behavior and if the effects were related to whether the user was programming or preparing a document. The Markov model which they developed has 16 states:

1. type one line
2. type many lines

3. change position by one line
4. change position by many lines
5. delete one line
6. delete many lines
7. insert one line
8. insert many lines
9. change position by one character
10. change position by many characters
11. delete one character
12. delete many characters
13. insert one character
14. insert many characters
15. search
16. search and replace

With this model, they studied the use of two editors (i.e., TECO and SOS) on a DEC System-10 and, using the methods of Anderson and Goodman, found that transition probabilities, averaged across individuals, were significantly affected by whether the user was programming or preparing a document and by the particular editor the user employed. However, when considering the distribution of individual transition matrices rather than the average matrix, they found that individual differences were so large that differences in tasks and/or editors were submerged. Thus, although a particular editor might have been better or worse for a group as a whole, they could not conclude that a particular editor was better or worse for specific individuals. These results led Hammer and Rouse to consider modeling user behavior at a level higher than the simple command level. In other words, they considered what the user was specifically doing rather than just analyzing the context-free sequence of editor commands. Their goal was to partition individual differences into behavioral and task differences.

Summary

It is interesting to contrast the three types of model presented in this chapter. Production systems are viewed as process models rather than statistical models in the sense that they describe the process by which behavior is exhibited rather than the probability that it will be exhibited or the distribution of times until it is exhibited (Gregg and Simon 1967). Although employing such models does allow one to avoid weak statistical statements such as accepting null hypotheses, it can present validation and generalization problems because individual differences must be dealt with rather than averaged out.

Further, when considering production systems as rank-ordered sets of pattern-evoked rules of thumb, one becomes concerned with how patterns are detected. This leads to pattern recognition models. Although some types of pattern recognition problem reasonably can be viewed as nonstatistical [e.g., the blocks world of Winston (1975)], many tasks are inherently probabilistic. Failure detection is an excellent example of such a situation. When probabilistic considerations become important, then the pattern recognition aspect of production system models causes the overall model to be a blend of process and statistical models.

When we begin to consider concatenations of probabilistically evoked productions, Markov chain models become of interest. Further, Markov models allow us to describe the behavior of populations of humans rather than just individuals. Although such aggregations certainly result in a loss of information, they do allow us to consider practical engineering questions such as, for example, designing a text editor for particular populations of users.

In this chapter, we have rather briefly reviewed production systems, pattern recognition, and Markov chains. The space devoted to each of these three methods has been considerably less than that devoted to the theories of estimation, control, queueing, and fuzzy sets. This choice of coverage does not imply that these three methods are less useful in a general sense. Instead, this choice reflects the engineering orientation of this book and the subsequent desire to devote more space to those methods that have already proven themselves in realistic human–machine contexts. It is quite possible that the methods discussed in this chapter will eventually become ubiquitous tools in systems engineering of human–machine systems. For that reason, this chapter has been devoted to briefly introducing them.

Chapter 7

Human–Machine Interaction in Retrospect

Throughout this book we have considered the problem of mathematically representing human–machine interaction. In pursuit of this problem, we have discussed a wide variety of mathematical methodologies drawn from engineering, computer science, management science, and mathematical psychology. As example applications, we have considered manual control of various vehicles, design of displays, detection and diagnosis of system failures, and many other tasks. In this chapter, we shall attempt to provide a rather broad perspective of human–machine interaction.

One approach to developing such a perspective might be to construct a taxonomy of modeling methodologies. Such a taxonomy would categorize methodologies in terms of appropriate task situations, independent variables, dependent variables, and performance measures. Although a taxonomy of this nature would probably be very useful, it would also be rather dry and mechanical. Further, it might serve to emphasize the methodology rather than the problem (i.e., human–machine interaction). For this reason, we shall instead consider a taxonomy of analogies.

Analogies Revisited

We have tried to repeatedly return to two basic concepts throughout this book: the use of analogies and the notion of constrained optimality. We view the first of these concepts (i.e., the use of analogies) as one of the most powerful tools of science and engineering. The second concept (i.e., constrained optimality) is fundamental to our ability to infer the impact of psychological and/or physiological constraints on human performance in other than a purely empirical manner.

At this point, we want to discuss a wide range of analogies of human behavior and consider how they can be useful in analyzing human–machine systems. We shall consider the following analogies:

1. Electrochemical network
2. Information processor
3. Pattern recognizer
4. Ideal observer
5. Servomechanism
6. Time-shared computer
7. Logical problem solver
8. Planner
9. Reflector/daydreamer

Notice that analogies 2–7 have been considered in previous chapters of this book whereas analogies 1, 8, and 9 are new. Notice also that the above list is arranged in ascending order according to a somewhat vague notion of increasing levels of conscious cognition.

At the lowest level of our taxonomy, we can view the human as a central nervous system involving electrochemical interactions among neurons. This would lead us to neural net models and sets of differential equations (Scott 1977). Although such models are fascinating, it is quite difficult to imagine them as useful for engineering design. Nevertheless, these models may prove useful for biomedical applications.

At the highest level of our taxonomy, the human can be viewed as a thinker in a broad, sketchy sense which we have chosen to connote with the words "reflecting" and "daydreaming." Although reflector/daydreamer is not really an analogy in the sense that the other entries on the above list are, it is included to represent a type of upper level or limit on our considerations of modeling. Thus, as with the electrochemical network analogy, the reflector/daydreamer analogy is viewed as not particularly useful in engineering design.[1]

We see that the analogies at the top and bottom of our list are included only for the purpose of delimiting the scope of our discussions. Therefore, we can loosely claim that our interests start above the physiological level and end below the creative level. Alternatively, or perhaps more appropriately, we should note that our expertise is limited to that range of interests.

---

[1] Of course, if one were trying to model the designer, as opposed to having the designer develop a model, then it is quite likely that reflecting/daydreaming should be viewed as an important part of the design process. However, we prefer to avoid the topic of modeling modelers, at least in this book.

The human can be viewed as an information processor in two ways. Newell and Simon (1972) describe the human in terms of processing symbols in short-term memory and storing/retrieving symbols from long-term memory. Although it would seem that such a low-level perspective would only be of practical applicability for simple situations such as reaction time tasks, Newell and Simon's indefatigable investigations have applied their information processing analogy to analysis of such robust tasks as playing chess. Nevertheless, as discussed in Chapter 6, it is quite cumbersome to consider realistic engineering design problems in terms of symbol manipulations in short-term and long-term memory. Thus, Newell and Simon's approach, at least in its original form, is best viewed as the basis of a behavioral theory rather than as an engineering design tool. (However, as the examples presented in Chapter 6 aptly illustrate, the modeling notions of Newell and Simon have been successfully extended for use in realistic engineering problems.)

The human can also be viewed as an information processor in the sense of being a communications channel. Employing information theory, the human can be represented as receiving bits (i.e., binary digits) of information which are then reduced or transformed in the process of producing an action or a decision. This notion of a communications channel allows us to describe the human in terms of channel capacity, noise characteristics, and so on.

As noted in Chapter 6, this form of the information processing analogy has received considerable attention. It appears to be quite appropriate in situations where a priori probabilities of the possible messages (i.e., inputs to the communications channel) are known. This constraint has resulted in applications of information theory being limited to laboratory investigations and other highly simplified task domains.

At a somewhat higher level in our taxonomy, we can view the human as a pattern recognizer. It is at this level that we can begin to consider realistic human–machine problems. Pattern recognition tasks are ubiquitous. Product inspection and process monitoring are excellent examples. In fact, there are many situations where the human is included in the system primarily because of pattern recognition abilities which often surpass any available algorithms. This presents some modeling problems because it is difficult to view the human as constrained optimal when the human can outperform the "optimal" algorithm. For this reason, as illustrated in Chapter 6, it is sometimes useful to construct pattern recognition algorithms on the basis of human behavior.

The ideal observer analogy fits roughly on the same level of our taxonomy as the pattern recognition analogy. Two versions of the ideal observer analogy deserve mention. First, there is the ideal observer based on estimation theory. Here we assume that the time series being observed by the human has some underlying structure in the sense that a correlation exists among elements of the series. Further, to employ this approach, we must have a knowledge of the underlying structure or at least be able to identify it. This assumption implies a tacit supposition that the human also has a perhaps unconscious knowledge of the structure of the process being observed. Given these assumptions, which are far from weak, we are then in a position to employ some rather powerful estimation algorithms as models of human behavior. We considered two applications of these algorithms in Chapter 2.

There is also an ideal observer analogy based on signal detection theory (Sheridan and Ferrell 1974, Chap. 19). The typical situation where this model has been applied involves the observation of a time series that may or may not contain a target signal. The human must respond "yes" or "no" concerning the presence of this signal. It is assumed that this decision is based on knowledge of a likelihood ratio L and knowledge of the conditional probability density functions for L given a signal and for L given no signal. Further, of course, the values of responding correctly and costs of incorrect responses are also included.

Although signal detection theory formulations have been frequently used, these applications have mainly occurred in the laboratory where the various probability density functions, values, and costs could be controlled. When we consider more realistic human—machine situations, it becomes more difficult to employ this approach. The main difficulty seems to be in determining how the human's observations produce a particular value of likelihood ratio L. One approach to solving this problem is by combining pattern recognition methods with signal detection theory (Greenstein and Rouse 1978).

Moving somewhat higher in our taxonomy of analogies, we come to the servomechanism analogy. Here the human must not only monitor the machine, but also must produce control signals that serve to guide the operaton of the machine. If viewed quite broadly, the servomechanism analogy has a very wide range of applicability. We shall consider this broad perspective later in this chapter. On the other hand, if the servomechanism analogy causes us to view the human as a simple error nulling device that continuously responds to errors in a deterministic and mechanical manner, then the range of applicability of this analogy is limited to such applications as vehicle control, process control, and artillery tracking. Nevertheless, these applications are very important and we, by no means, intend to denigrate work in these areas. Indeed, many of the relatively few human—machine systems for which there are excellent models are in these areas.

Considering a rule of thumb to determine how strict an interpretation of the servomechanism analogy applies, the following generalization comes to mind. When the response time of a machine is relatively fast (e.g., lateral control of automobiles and bicycles), a continuously controlling servomechanism is a reasonable model. However, as the response time decreases (e.g., jumbo jet, supertanker, nuclear power plant, and inventory control system), the continuously controlling servomechanism analogy becomes increasingly inappropriate. As this occurs, one must either change analogies or broaden one's view of servomechanisms. At this point, we shall pursue the former alternative.

The next level in our taxonomy of analogies includes the time-shared computer. This analogy is particularly fascinating because the many rapid advances in computer technology have led to corresponding rapid refinements in our view of the human as a time-shared computer. This analogy is primarily of use when the human has responsibility for multiple tasks. First, the human must detect that tasks are awaiting attention; resources (usually time) are then allocated in order to meet task demands. The time-shared computer analogy is most useful when the multiple tasks differ in terms of when they demand attention, how much attention is demanded, and the value of performing tasks or cost of delaying tasks. On the other hand, if all tasks are the same in terms of these characteristics, the time-shared computer analogy is still applicable but certainly not as interesting.

Of all the analogies that we have considered thus far, the time-shared computer is perhaps the first that realistically captures the essence of the human's involvement in tasks such as driving automobiles and piloting aircraft. It allows us to represent behavior in terms of how the human deals with dissimilar tasks in order to achieve overall system goals. In fact, if one stretches it a bit, much of our everyday activity can be represented using a time-shared computer analogy. For example, most of us must share our time between meeting with colleagues, answering phone calls, writing letters, working on various projects, and so on. Each of these activities have differing frequencies of occurrence, time requirements, and values or costs.

Thus, the time-shared computer analogy appears to have wide applicability. However, it does have definite limitations. Although models based on this analogy can tell us how long tasks will typically wait for attention and how busy the human will be, these models cannot tell us too much about how well a task is performed. Therefore, for example, a model based on a time-shared computer analogy might indicate that a human in a particular task will spend an average of 20% of the time scanning status displays. However, this model would say little about what the human sees on those displays.

We now come to the logical problem solver analogy. In some ways, it may not be clear why logical problem solver qualifies as an analogy. It is not particularly associated with a piece of equipment such as a servomechanism or a computer. However, it is premised on an idealized notion that problem solving is performed using various set theory operations. Thus, we are viewing the human as a machine that manipulates sets. Just as with pattern recognition or estimation theory, the human is assumed to formulate problems in a particular manner and then the logical implications of that formulation are pursued and compared to actual human behavior.

Although human problem solving has received considerable attention within psychology, it has only recently become of interest within engineering design. The motivation is quite clear. As the human's role increasingly becomes one of taking over from automation in failure situations, systems designers need methods of incorporating human problem solving abilities into the design equations. Although, as discussed in Chapter 5, some progress in this area has been made, it does not seem realistic to hope for problem solving models with the degrees of both specificity and generality possible with, for example, servomechanism models. The difficulty is that highly cognitive tasks such as problem solving allow much wider variations in strategy and time–accuracy trade-offs than possible with control tasks that are dominated by requirements to stabilize the dynamics of the human–machine system.

The planner is the last of the relatively basic analogies that we shall consider. Subsequently, we shall discuss composite analogies whose elements are basic in the sense of reflecting a particular, somewhat narrow perspective of human behavior. Planning offers an interesting transition from basic to composite analogies because, as we shall later discuss, it is difficult to classify planning as either basic or composite.

The Human as Planner

Johannsen and Rouse (1979) informally collected written protocols of two automobile drivers explaining what they would be doing as they went on a hypothetical automobile trip. Studying these protocols, it was readily apparent that many of the individuals' conscious activities were devoted to developing, initiating, and monitoring plans. This observation led Johannsen and Rouse to consider various notions about planning. We shall review their conclusions here.

The human develops a plan in hopes that its execution will achieve some goals. Although in most human–machine systems one usually accepts the overall goal as given (e.g., land the aircraft), the process of developing subgoals is often left to the human. The partitioning of goals into subgoals and then subgoals into lesser subgoals, and so on, reflects a hierarchical mode of planning that has received considerable attention (Sacerdoti 1975; Weissman 1976).

The hierarchical approach allows one to develop plans that are broad and sketchy as opposed to detailed and concise. Thus, low-level subgoals can be temporarily ignored until their immediacy demands attention. Similarly, future actions which require preconditions that are not as yet assured can perhaps be temporarily ignored if one feels that the environment is "hospitable" to one's goals (Weissman 1976). In this way, the human can delay the planning process and, in many cases, avoid investing effort in developing contingency plans that are unlikely to be needed.

On the other hand, low-level subgoals eventually demand attention and must be dealt with. Then, a concise system dynamics model such as Carbonell's probably provides a reasonable description of human behavior (Carbonell 1969). This model assumes that the human is dealing with a system describable by quantitative state transitions and amenable to quantitative control actions. Thus, this model fits into the ideal observer/ servomechanism categories discussed in Chapters 2 and 3.

Such low-level planning is probably unconscious in the sense that the human is unaware of it. From the perspective of a computer analogy, one might say that high-level, conscious planning is like executing an interpreted program. (An interpreted program is one in which the computer "consciously" has to interpret the meaning of each statement as it is executed.) On the other hand, low-level unconscious planning is similar to executing a compiled program (Newell and Simon 1972). In fact, it might be claimed that low-level planning cannot really be called planning. Instead, such activities are only the details of implementation which are carried out by various pattern recognizers, ideal observers, and servomechanisms.

Planning appears to include the following six aspects:

1. Generation of alternative plans
2. Imagining of consequences
3. Valuing of consequences
4. Choosing and initiating a plan
5. Monitoring plan execution
6. Debugging and updating plan

The last two aspects deal with observing plan execution and subsequent replanning rather than with actual implementation.

How might one model the generation of alternative plans? One can look at a plan as a linked set of subplans (Weissman 1976). However, at some level, subplans must be specific. In many tasks, the alternatives are clearly defined at the outset. On the other hand, there are many interesting tasks (e.g., dealing with totally unanticipated system failures) where the human must create alternatives. In such cases, humans usually first consider alternatives that have been successful in previous situations.

One might use Newell's pattern-evoked production systems (see Chapter 6) as a model of how the human accomplishes this search for alternatives. In other words, one could view alternative plans as being evoked by the "pattern" of the subgoals and environment. As an alternative to production systems, the idea of scripts might provide a reasonable model. "A script is a structure that describes appropriate sequences of events in a particular context" (Schank and Abelson 1977). [Minsky's notation of "frames" is somewhat similar to this idea of scripts (Minsky 1975).] In other words, a script is somewhat like a standard plan or subplan that is evoked in particular situations. For example, many people probably have "driving to work" scripts which they unconsciously, but nevertheless faithfully, follow.

The ideas of production systems and scripts are both related to the idea of the human having an internal model. However, a production system or script is very different from the type of model assumed in the system dynamics domain. Namely, production systems and scripts provide forecasts of typical consequences rather than models of internal state transitions. This type of forecast is accomplished by a direct mapping rather than by any process of calculating.

Sometimes a *new* alternative is needed and it is very difficult to say how a totally new idea is generated. Linking the idea of associative memory (Anderson and Bower 1973; Kohonen 1977) with the idea of production systems or scripts, one can conjecture that new ideas are generated when the criterion for matching the new subgoal with past experiences is relaxed and/or nonstandard features of the situation are emphasized. At this level of planning, we are approaching the aforementioned domain of reflecting/daydreaming.

Long-term plans that will not be immediately implemented are probably developed at the highest level in the goal hierarchy with only major goals considered. Such a plan might be a somewhat vague verbal statement or perhaps a sketch of activities and relationships. It is interesting to speculate upon (and perhaps research) what plans look like in the "mind's eye." For example, are plans listlike or are they more spatial, such as Warfield's (1976) interpretive structural models? Rasmussen (1979) provides a rather broad review of alternative ways in which the human might internally represent various phenomena.

Short-term plans that will require immediate implementation cannot be quite so sketchy. In this case, the human has to consider specific actions. As noted earlier, one would probably be reasonably successful in modeling this type of plan using something like Carbonell's model or a formulation employing production systems where specific features of the environment would automatically evoke particular responses.

Given a set of candidate plans, the human must forecast or imagine the consequences of implementing each plan. One might assume that the human performs some type of mental simulation of the plan. For example, the human might use the current perception of the system dynamics to extrapolate the system's state as a function of planned control strategy. The model of human behavior in estimation tasks that was presented in Chapter 2 could be employed to represent this type of forecasting.

However, when plans are sketchy, at least in terms of intermediate preconditions, the human probably does not actually calculate consequences but instead simply maps plan features to previously experienced consequences. Then, until evidence overrides the assumption, the human assumes these previously experienced consequences will prevail. This type of behavior is represented quite nicely by the scripts concept (Schank and Abelson 1977).

Imagined consequences are then compared to goals. For low-level plans, the comparison might be based on a well-defined criterion function. However, this is probably not the case for high-level plans. Since high-level goals and imagined consequences may be verbal and rather vague, it is likely that the human only tries to "satisfice" rather than optimize, that is, searches for satisfactory rather than optimal solutions. One might represent this phenomenon using multi-attribute utility functions (Keeney and Raiffa 1976) that have broad optima. Alternatively, concepts from fuzzy set theory (see Chapter 5) might be used to consider the membership of a set of consequences in the fuzzy set of acceptable consequences. The utility function approach is probably appropriate if one assumes that the human has a fairly precise knowledge of the possible consequences, and subsequently values some more than others. On the other hand, the fuzzy set approach would seem to be applicable to situations where the human's perception of the consequences is actually fuzzy.

The human chooses the most satisfactory plan and initiates its execution. If none of the available plans meets an acceptable level of satisfaction, the human either tries to debug the set of plans under consideration or perhaps tries to develop new plans. Debugging of partially failed plans may initially involve local experimentation to determine the cause of plan failure rather than a global reevaluation and complete replanning (Davis 1977). One approach to modeling the debugging or troubleshooting of plans is the fault diagnosis model discussed in Chapter 5.

Assuming that a plan has been initiated, the human monitors its execution and only becomes involved (in the sense of planning) if the unanticipated occurs or execution reaches the point that some phase of the plan must be more concisely defined. Monitoring for the unexpected might be modeled using production systems that trigger when the preconditions are *not* satisfied. Other approaches, based on estimation theory and pattern recognition methods were discussed in Chapters 2 and 6, respectively.

Once the unexpected has been detected, planning might shift into the above-mentioned debugging mode. On the other hand, the need to shift from sketchy to concise planning may involve abandoning, for the moment, the broad hierarchical mode and shifting to a detailed partially preprogrammed mode.

How do all these bits and pieces fit into an overall view of planning? It does seem that the hierarchical approach to planning combined with the production system and script ideas provide a reasonable framework. This framework offers a plausible explanation for how the human manages to cope with complex environments. However, one might quite rightly question the notion that the hierarchical planner fulfills the definition of an analogy. Perhaps it would be better to simply view planning as a structure within which various of our other analogies can be embedded?

Composite Analogies

A composite analogy is a structure or framework within which basic analogies can be incorporated. As it has a structure, the composite analogy is more than just the sum of its components. For example, our discussion of the human as a planner included consideration of the human as an information processor, ideal observer, servomechanism, time-shared computer, and logical problem solver. However, the essence of planning was viewed as the hierarchical manner of dealing with uncertain environments. Thus, the structure of a composite analogy serves to integrate the basic analogies into a coherent purposeful entity.

It appears that composite analogies emerge once we begin to consider realistic human—machine systems. Perhaps the simplest composite is the optimal control model (discussed in Chapter 3) which includes both the ideal observer and servomechanism. Carbonell's conceptual model of human—computer interaction offers a somewhat similar composite (Carbonell 1969). In these two models, the way in which ideal observer and servomechanism interact is far from arbitrary and, in fact, very important. Thus, as we noted earlier, the structure of the composite is quite significant.

In Chapter 3, we briefly discussed Sheridan's notion of supervisory control. In considering the role of the human as a supervisor of automatically controlled processes, Sheridan (1976) discussed four modes of supervision: planning, teaching, monitoring, and intervening. While this conceptual model has not been pursued in great depth, it seems reasonable to suggest that such elaboration would result in a composite analogy incorporating virtually every topic discussed in this book. Of course, quantitatively manipulating such a collage of analogies would present difficulties. In fact, as we attempt to gather more and more of human behavior within a composite analogy, there is usually less and less that we can quantitatively conclude. Nevertheless, such exercises can prove to be quite interesting.

For example, it is interesting to consider how the various analogies discussed here might be useful for describing the behavior of groups of individuals. Although this topic is beyond the scope of this book and, when viewed too broadly, is subject to much criticism (Lilienfeld 1978), one composite analogy is worth noting. When viewing a particular individual within a hierarchical organization of multiple individuals, one might wonder what basic analogy would best describe that particular individual's role in the overall organization. The answer depends on where the observer sits in the hierarchy. If it is such that the individual being observed is lower in the hierarchy than the observer, then one would probably conclude that the individual is best described as a servomechanism. On the other hand, if the individual being observed is higher in the hierarchy than the observer, then it is likely that one would choose to represent the individual as a time-shared computer. Thus, we have a composite analogy that views organizations as hierarchies of servomechanisms receiving commands from time-shared computers. Further, our composite describes each individual in the organization as viewing subordinates as servomechanisms and superiors as time-shared computers. We hasten to note that this composite analogy of organizations is offered merely because it is interesting and not because it has proven to be particularly useful.

In the past few paragraphs, we have considered planning and supervisory control at some length as well as organizational decision making quite briefly. At this point, the reader may wonder if we perhaps have reached a level of generality at which the spirit of mathematical modeling espoused in this book no longer applies. To consider this issue more explicitly, we shall now discuss what appear to be some limitations of the modeling approach.

Limitations of Models

There are three types of limit that affect one's ability to develop practically useful models of human–machine interaction successfully:

1. Measurement and computational difficulties
2. Nonuniqueness of constrained optimality
3. Pervasiveness of task environments

Measurement problems arise when one has to make inferences about processes within the human, the structure and parameters of which cannot be directly observed. Alternatively, if at least indirect observation is possible, the measures of interest may be quite noisy due to the nonrepeatable and time-varying nature of human behavior as well as individual differences. In other words, in many realistic tasks, no two individuals will perform exactly alike and further, the same individual performing the task at two different times is unlikely to produce exactly the same behavior twice.

Although we could develop models for particular individuals at particular times, such models are not very useful from an engineering design point of view. Thus we resort to modeling average behavior. Unfortunately, we lose information as well as noise in the averaging process. However, since we usually cannot take the human apart for the purpose of observing the internal mechanisms, we have to accept this limitation. Indeed, even if we were allowed to dissect people, it would be quite difficult to find the essence of decision making, problem solving, or planning among the various products of the dissection process. Thus, we end up modeling aggregate behavior when, in fact, behavior at particular instances may be quite interesting (e.g., during the onset of emergencies preceded by specific states of mind).

As we attempt to model increasingly complex human—machine systems, our models often tend to become more and more elaborate. For example, pattern recognition in unconstrained environments would require a rather complicated model. Computational difficulties arise when we try to use such models to compute behavioral patterns and resulting performance. It is not unusual for very sophisticated computers to require hours to solve a pattern recognition and interpretation problem (e.g., reading) that a human could solve in a matter of seconds. Although we might hope for faster and bigger computers, it seems reasonable to claim that in areas such as vision and language the computational problems may be almost insurmountable, at least with our current ways of looking at these domains of human behavior.

However, we need not wander off into vision and language to encounter computational difficulties. Some of the models reviewed in this book can present storage and computation time problems when applied to realistically complex tasks. Also, of course, storage and time can affect the cost of using a model. Thus the usefulness of a model may be limited by the cost of using it as compared to the benefits derived. Although this issue may not affect those researchers who have essentially free computer services, it may be of interest to a systems designer.

Throughout our discussions of various models, we have stressed the usefulness of the notion of constrained optimality. Using this idea allows one to formulate human—machine interaction as an optimization problem. In this way, a wide variety of engineering tools can be employed to solve the problem (i.e., determine the optimal behavior). Unfortunately, this powerful approach presents two difficulties. First of all, it is not clear that the human is actually an optimizer. As we briefly noted during our discussion of planning, it is probably more appropriate to view the human as a "satisficer." In other words, one does not always "do one's best" but instead only attempts to find satisfactory solutions to problems. In many tasks, this approach enables the human to avoid worrying about details until it becomes necessary to do so. This can result in a drastic reduction in work load and also allow the human to retain the resources to deal with more tasks as well as flexibly react to unforeseen events. These characteristics are precisely the reasons why humans are often included in systems.

But how do we model satisficing behavior? Although we could discard optimization, this would leave us stripped of one of our most important tools and without a viable alternative. Instead, we can look at optimization with respect to criteria that allow multiple satisfactory solutions. Then, in comparing human behavior to that of the model, success would only require that the human choose one of the set of equally attractive decisions. Although this approach solves the problem of satisfaction versus optimization, it does have important limits. For example, in multistage decision-making tasks, the environment at state i depends on the decisions made at stages $i - 1$, $i - 2$, and so on. If a number of satisfactory solutions are possible at each stage, then one can see that there will be many possible satisfactory decision sequences, and the model and human will only agree if they both happen to choose the same sequence. Further, if at any stage the model and human disagree, then from that point on they both are working on different problems. This can present considerable difficulty in terms of validating a model. Thus, when using such a model as a predictive tool, one can only say that the human will choose a decision sequence that is equally satisfactory to what the model chose and, with a small probability, may be the same sequence chosen by the model.

Thus, we have a problem of nonuniqueness of the solution to our model formulation. There is also another source of nonuniqueness. We have noted throughout our discussions that constraints usually must be added to the optimization problem if the model is to perform in a manner similar to that of the human. Therefore, we considered reaction time delay, neuromotor sluggishness, observation noise, motor noise, limited memory, and so on. The difficulty with these constraints is that each does not produce unique results. Thus, for example, one can achieve the same results by assuming observation noise or by assuming an approximate internal model of the process dynamics. This problem can be most troublesome when one is using noise sources as constraints (as opposed to structural constraints) while comparing model and human only on the basis of some global performance measure such as root-mean-squared error. The most important limit imposed by this type of nonuniqueness involves the degree to which one can make realistic behavioral interpretations of the model's performance. Obviously, when the model is only one of several equally accurate alternatives, one's behavioral interpetatons must include a few caveats. One of the main ways of resolving this difficulty is through repeated experimentation with a multiplicity of tasks.

Simon has emphasized the extent to which the environment can be quite pervasive within any attempt to model human behavior (Simon 1969). He is quite convincing in claiming that human behavior mainly reflects the task environment. Thus searching for a specific analytical model of general human behavior may only be fruitful to the extent that all task environments are common. Perhaps then one should first search for commonality among environments rather than intrinsic human characteristics. In other words, a good model of the demands of the environment may allow a fairly accurate prediction of human performance, because it is reasonable to assume, at least initially, that the human will adapt to the demands of the task and perform accordingly.

From this perspective, we perhaps are developing models of machines rather than models of humans. However, things are really not that simple. The way in which humans respond to an environment depends on how they understand that environment (Rouse and Rouse 1979). Further, it depends on the human's abilities to sense environmental stimuli and produce desired actions. Therefore, we have to know quite a bit about the human as well as the environment to be able to model human–machine interaction successfully. Nevertheless, Simon is certainly correct to the extent that it is very difficult to say much about human behavior in the absence of any context. For this reason, the usefulness of many of our models is context dependent. When we try to avoid specific contexts, our models often become qualitative glittering generalities which typically are not particularly useful for engineering design.

Two other limitations deserve to be mentioned briefly. In our discussion of planning, we noted that activities such as reflecting and daydreaming filled the interstices of the planning process. Although a *total* model of human behavior should capture such activities, not to mention even higher-level activities, we really cannot hope to formulate such phenomena mathematically. There are many types of behavior that are simply beyond the reach of our tools. Thus we are limited, perhaps fortunately, from increasingly capturing more of the human in our equations.

A final limitation involves a basic difficulty in many domains where mathematical models are applied. Namely, it often is the case that the only person who totally understands a model is the person who developed it. In other words, to understand a model to the extent that one comprehends the philosophy on which it is based, its limitations, and how one can most successfully manipulate it, one usually has to have been involved with the development of the model. This is particularly true of new models. After models have received years of attention this problem tends to disappear. Thus we sometimes find new models being misapplied, at least to the extent that the results are overinterpreted.

After considering all of the above limitations, is the modeling process still credible? The answer is "yes" and, in fact, credibility is increased by recognizing these limitations and then plying our craft within them. In this way, we shall not expect more than we can produce and we shall be more successful in producing what we expect.

Future Prospects

It seems reasonable to argue that human–machine systems will increasingly become human–computer systems with the computer taking over much of the continuous control tasks and, perhaps eventually, many of the pattern recognition tasks. Further, computers will become increasingly "intelligent" in the sense that they will be able to perform decision-making tasks that once were thought to be necessarily the responsibilities of the human. This possibility raises the question of how humans and computers are going to communicate.

Although there are a variety of devices suitable for human–machine communications, ranging from keyboards and light pens to natural language (Rouse 1975; Rijnsdorp and Rouse 1977), they are not sufficient to solve the communication problem completely. What is necessary is a means for the computer to "understand" the human. One approach to giving the computer this understanding is to embed within its software models of human behavior (Rouse 1975, 1977b). With such models the computer may be able to put the human's queries and responses into some structure and, in that way, interact with the human in a manner more like human–human interaction.

Although this concept is definitely a future prospect rather than a current reality, several of the models discussed in earlier chapters are initial attempts in this direction. For example, the queueing model of flight management, the Markov model of editing behavior, and the fuzzy set model of fault diagnosis were, in varying degrees, developed with human–computer interaction in mind. Thus, the idea of embedding models in computers is not purely conjecture.

As the human becomes less of an active controller and more of a supervisor, an increasing proportion of the time will be spent performing higher-level tasks. Planning is an excellent example of such tasks. This trend is going to cause us to ask new questions. For example, instead of asking how a particular display will affect control performance, we shall be asking how alternative displays will affect the human's ability to plan. Thus, we shall need models of human–machine interaction in planning tasks. This will pose many problems.

One problem will be observing the human's planning process. It would seem that either verbal or written protocols will be necessary and hopefully useful (Newell and Simon 1972; Rasmussen and Jensen 1974; Johannsen and Rouse 1979). However, collecting and analyzing such protocols can be extremely time consuming. What is needed is an automated method of accomplishing this task. Some efforts have been invested in this direction (Bhaskar and Simon 1977). Of course, even if these efforts are successful, we are still faced with the problem that verbal and written protocols tend to be rather vague and qualitative. Of the various modeling methodologies discussed in this book, only fuzzy set theory, production systems, and pattern recognition are directly suitable for such nonquantitative analyses. However, although we do have a few tools with which to attack these problems, we should not expect the quantitative success that has been possible in such domains as manual control.

Another role for the computer in future human–machine systems is that of intelligent assistant in the sense that it will not only carry out your instructions but also instruct you how the system works, detect when your performance is degrading, and perhaps take over in some caretaking mode if all else fails. To realize all of these possibilities, the computer will need more than just models of how the human successfully performs tasks. It will also need models of human behavior when the human is naive, making mistakes, and so on. This presents an interesting challenge because we shall no longer be able to obtain so many clues from the environment. We shall have to understand the possible ways in which a human can misunderstand a problem.

In Chapter 1, we suggested that the servomechanism is the only analogy within human–machine systems research that has achieved the status of a paradigm as defined by Kuhn (1962). However, thoughout this book the servomechanism analogy has only served as an equal among many analogies. Further, in Chapter 3 on control theory, we alluded to the possibility that the usefulness of the servomechanism analogy may reach a limit as the topic of supervisory control is pursued. Does this mean that we expect a new paradigm to emerge within human–machine systems research to replace the servomechanism?

In fact, we should like to suggest quite a different possibility. We noted that an alternative expression for the servomechanism analogy is the cybernetic analogy. It is unfortunate that these two phrases have come to be synonymous. The servomechanism as an error-nulling device may be an appropriate description of some aspects of automobile driving and aircraft piloting, but in general the servomechanism has come to represent a somewhat narrow view of the concept of cybernetics. It seems that it is an appropriate time to broaden our perspective.

Thus we propose that the human be viewed as an organism who receives inputs from the environment, compares the inputs to what was expected, processes these two types of information in a variety of ways, and then perhaps, but not necessarily, produces some action that may modify the environment. This process continually iterates and thereby people walk, drive automobiles, repair airplane engines, manage insurance companies, and so on. Within this general framework, we can easily place all of the analogies and methodologies discussed in this book. This leads one to view the problem of control as much broader than the current field of control theory. For example, using queueing theory to represent and solve a problem involving control of a time-shared computer (human or otherwise) is a control problem despite the fact that control theory is not employed in its solution. As another example, we can view problem solving and planning as ingredients in the development of control strategies, even though it is difficult to represent these activities within a conventional control theoretic formulation.

Therefore, as a final future prospect, we suggest that the concept of control and the cybernetic paradigm should be viewed quite broadly, at least in order to remain relevant, but also to allow integration of the various emerging modeling methodologies within a coherent framework. The availability of such a framework will allow us to retain the intellectually satisfying view of humans as organisms whose central problem is control of themselves and of the environment in order to achieve some overall goals.

References

Afifi, A. A., and Azen, S. P. 1972. *Statistical Analysis*. New York: Academic Press.

Allen, A. O. 1975. Elements of Queueing Theory for System Design. *IBM Systems Journal* 14 (No. 2), 161–187.

Anderson, B. D. O., and Moore, J. B. 1979. *Optimal Filtering*. Englewood Cliffs, NJ: Prentice-Hall.

Anderson, J. R., and Bower, G. H. 1973. *Human Associative Memory*. New York: Wiley.

Anderson, T. W., and Goodman, L. A. 1957. Statistical Inference about Markov Chains. *Annals of Mathematical Statistics* 28, 89–110.

Astrom, K. J., and Eykhoff, P. 1971. System Identification—A Survey. *Automatica* 7 (No. 2), 123–162.

Athans, M. 1971. The Role and Use of the Stochastic Linear-Quadratic-Gaussian Problem in Control System Design. *IEEE Transactions on Automatic Control* AC-16 (No. 6), 529–552.

Attneave, F. 1959. *Applications of Information Theory to Psychology*. New York: Holt.

Bar-Shalom, Y. 1972. Optimal Simultaneous Estimation and Parameter Identification in Linear Discrete-Time Systems. *IEEE Transactions on Automatic Control* AC-17 (No. 3), 308–319.

Baron, S. 1977. Some Comments on Parameter Identification in the Optimal Control Model. *Systems, Man, and Cybernetics Review* 1, 4–6.

Baron, S., and Levison, W. H. 1977. Display Analysis Using the Optimal Control Model of the Human Operator. *Human Factors* 19 (No. 4), 437–457.

Bell, C. E. 1973. Optimal Operation of an M/G/1 Priority Queue with Removable Server. *Operations Research* 21 (No. 6), 1281–1290.

Bhaskar, R. and Simon, H. A. 1977. Problem Solving in Semantically Rich Domains: An Example from Engineering Thermodynamics. *Cognitive Science* 1 (No. 2) 193–215.

Billingsley, P. 1961. *Statistical Inference for Markov Processes.* Chicago: University of Chicago Press.

Birmingham, H. P., and Taylor, F. V. 1954. A Design Philosophy for Man–Machine Control Systems. *Proceedings of the IRE* 42, 1748–1758.

Carbonell, J. R. 1966. A Queueing Model of Many-Instrument Visual Sampling. *IEEE Transactions on Human Factors in Electronics* HFE-4 (No. 4), 157–164.

Carbonell, J. R. 1969. On Man–Computer Interaction: A Model and Some Related Issues. *IEEE Transactions on Systems Science and Cybernetics* SSC-5 (No. 1), 16–26.

Carbonell, J. R., Ward, J. L., and Senders, J. W. 1968. A Queueing Model of Visual Sampling: Experimental Validation. *IEEE Transactions on Man–Machine Systems* MMS-9 (No. 3), 82–87.

Chu, Y. 1976. Optimal Control of Queues. *Seminars on Queueing Systems* (W. B. Rouse, Ed.). University of Illinois UILU-ENG-77-4002.

Chu, Y. 1978. Adaptive Allocation of Decision Making Responsibility Between Human and Computer in Multi-Task Situations. Ph.D. thesis, University of Illinois at Urbana-Champaign, 1978.

Chu, Y. Y., and Rouse, W. B. 1979. Adaptive Allocation of Decision Making Responsibility Between Human and Computer in Multi-Task Situations. *IEEE Transactions on Systems, Man, and Cybernetics* SMC-9 (No 12), 769–778.

Conant, R. C., and Ashby, W. R. 1970. Every Good Regulator of a System Must be a Model of That System, *International Journal of Systems Science* 1 (No. 2), 89–97.

Curry, R. E. 1971. A Bayesian Model for Visual Space Perception. *Proceedings of the Seventh Annual Conference on Manual Control*, University of Southern California. Moffett Field, CA: NASA, pp. 187–196.

Curry, R. E., and Govindaraj, T. 1977. The Human as a Detector of Changes in Variance and Bandwidth. *Proceedings of the Thirteenth Annual Conference on Manual Control*, MIT. Moffett Field, CA: NASA, pp. 217–221.

Curry, R. E., Kleinman, D. L., and Hoffman, W. C. 1977. A Design Procedure for Control/Display Systems. *Human Factors* 19 (No. 4), 421–436.

Davis, P. R. 1977. Using and Re-using Partial Plans. Ph.D. thesis, University of Illinois at Urbana-Champaign.

Dey, D. 1975. Problems, Questions and Results in the Use of the BBN Model. *Proceedings of the Eleventh Annual Conference on Manual Control*, NASA Ames Research Center. Moffett Field, CA: NASA #TM X-62, 464, pp. 577–588.

Donges, E. 1978. A Two-Level Model of Driver Steering Behavior. *Human Factors* 20 (No. 6), 691–707.

Edwards, E., and Lees, F. P. 1974. *The Human Operator in Process Control.* London: Taylor and Francis.

Enstrom, K. D. 1976. Real Time Adaptive Modeling of the Human Controller with Applications to Man-Computer Interaction, MSIE thesis, University of Illinois at Urbana-Champaign.

Enstrom, K. D., and Rouse, W. B., 1977. Real-Time Determination of How a Human Has Allocated His Attention Between Control and Monitoring Tasks. *IEEE Transactions on Systems, Man, and Cybernetics.* SMC-7 (No. 3), 153–161.

Estes, W. K., and Suppes, P. 1974. Foundations of Stimulus Sampling Theory. In *Contemporary Developments in Mathematical Psychology* (D. H. Krantz et al., Eds.). San Francisco: Freeman, pp. 163–183.

Eykhoff, P. 1974. *System Identification: Parameter and State Estimation.* New York: Wiley.

Ferrell, W. R., and Sheridan, T. B. 1967. Supervisory Control of Remote Manipulation. *IEEE Spectrum* 4 (No. 10), 81–88.

Fitzgerald, R. J. 1971. Divergence of the Kalman Filter. *IEEE Transactions on Automatic Control* AC-16 (No. 6), 736–747.

Gai, E. G., and Curry, R. E. 1976a. A Model of the Human Observer in Failure Detection Tasks. *IEEE Transactions on Systems, Man, and Cybernetics* SMC-6 (No. 2), 85–94.

Gai, E. G., and Curry, R. E. 1976b. Failure Detection by Pilots During Automatic Landings: Model and Experiment. *Journal of Aircraft* 14 (No. 2), 135–141.

Gaines, B. R., and Kohout, L. J. 1977. The Fuzzy Decade: A Bibliography of Fuzzy Systems and Closely Related Topics. *International Journal of Man–Machine Studies* 9 (No. 1), 1–68.

Goldstein, I. P., and Grimson, E. 1977. Annotated Production Systems: A Model for Skill Acquisition. *Proceedings of the Fifth International Joint Conference on Artificial Intelligence*, MIT. Cambidge, MA: MIT Artificial Intelligence Laboratory, pp. 311–317.

Govindaraj, T. 1979. Modeling the Human as a Controller in a Multi-Task Environment, Ph.D. thesis, University of Illinois at Urbana-Champaign.

Govindaraj, T., and Rouse, W. B. 1979. Modeling Human Decision Making in Multi-Task Situations Involving Both Control and Discrete Tasks. *Proceedings of the Fifteenth Annual Conference on Manual Control*, Wright State University. Wright Patterson AFB, OH: Air Force Flight Dynamics Laboratory.

Greeno, J. G. 1974. Representation of Learning as Discrete Transition in a Finite State Space. In *Contemporary Developments in Mathematical Psychology* (D. H. Krantz et al., Eds.). San Francisco: Freeman, pp. 1–43.

Greenstein, J. S. 1979. Human Decision Making in Multi-Task Situations: Event Detection, Attention Allocation, and Implications for Computer Aiding. Ph.D. thesis, University of Illinois at Urbana-Champaign.

Greenstein, J. S., and Rouse, W. B. 1978. A Model of Human Event Detection in Multiple Process Monitoring Situations. *Proceedings of the Fourteenth Annual Conference on Manual Control*, University of Southern California. Moffett Field, CA: NASA, Conf. Publ. 2060.

Gregg, L. W., and Simon, H. A. 1967. Process Models and Stochastic Theories of Simple Concept Formation. *Journal of Mathematical Psychology* 4, 246–276.

Hammer, J. M. and Rouse, W. B. 1979. Analysis and Modeling of Freeform Editing Behavior. *Proceedings of the 1979 International Conference on Cybernetics and Society*, Denver. New York: IEEE.

Heffes, H. 1966. The Effect of Erroneous Models on the Kalman Filter Response. *IEEE Transactions on Automatic Control* AC-11 (No. 3), 541–543.

Heyman, D. P. 1968. Optimal Operating Policy for M/G/1 Queueing Systems. *Operations Research* 16, 362-382.

Howard, R. A. 1960. *Dynamic Programming and Markov Processes*. Cambridge, MA: MIT Press.

Hunt, R. M. 1979 A Study of Transfer of Training from Context-Free to Context-Specific Fault Diagnostic Tasks. MSIE thesis, University of Illinois at Urbana-Champaign.

IEEE 1977. Special Issue on Current Perspectives in Pattern Recognition. *Systems, Man, and Cybernetics Review* 6 (No. 4).

Inooka, H., and Inoue, A. 1978. Application of the GMDH Algorithm to a Manual Control System. *IEEE Transactions on Systems, Man, and Cybernetics* SMC-8 (No. 11), 819–821.

Ivankhenko, A. G. 1971. Polynomial Theory of Complex Systems. *IEEE Transactions on Systems, Man, and Cybernetics* SMC-1 (No. 4), 364–378.

Johannsen, G., Boller, H. E., Donges, E., and Stein, W. 1977. *Der Mench im Regelkreis: Lineare Modelle*. Munchen: Oldenbourg.

Johannsen, G., and Govindaraj, T. 1980. Optimal Control Model Predictions of System Performance and Attention Allocation and the Experimental Validation in a Display Design Study. *IEEE Transactions on Systems, Man, and Cybernetics* SMC-10 (No. 5).

Johannsen, G., and Rouse, W. B. 1979. Mathematical Concepts for Modeling Human Behavior in Complex Man–Machine Systems. *Human Factors* 21 (No. 6).

Johnson, W. B. 1979. Computer Simulations in Fault Diagnosis Training: An Empirical Study of Learning Transfer from Simulation to Live System Performance. Ph.D. thesis proposal, University of Illinois at Urbana-Champaign.

Kaufman, A. 1975. *Introduction to the Theory of Fuzzy Subsets*. New York: Academic Press.

Keeney, R. L., and Raiffa, H. 1976. *Decision With Multiple Objectives*. New York: Wiley.

Kelley, C. R. 1968. *Manual and Automatic Control*. New York: Wiley.

King, P. J., and Mamdani, E. H. 1977. The Application of Fuzzy Control Systems to Industrial Processes. *Automatica* 13 (No. 3), 235–242.

Kleinman, D. L. 1975. Comments on the Application of Modern Control Theory to Manual Control. *Man–Machine Systems Review* 1, 12–13.

Kleinman, D. L. 1976. Solving the Optimal Attention Allocation Problem in Manual Control. *IEEE Transactions on Automatic Control* AC-21 (No. 6), 813–822.

Kleinman, D. L., Baron, S., and Levison, W. H. 1971. A Control Theoretic Approach to Manned-Vehicle System Analysis. *IEEE Transactions on Automatic Control*. AC-16 (No. 12), 824–832.

Kleinrock, L. 1975. *Queueing Systems*. New York: Wiley, Vol. I.

Kleinrock, L. 1976. *Queueing Systems*. New York: Wiley, Vol. II.

Kohonen, T. 1977. *Associative Memory*. New York: Springer-Verlag.

Kok, J. J., and Van Wijk, R. A. 1977. A Model of the Human Supervisor. *Proceedings of the Thirteenth Annual Conference on Manual Control*, MIT. Moffett Field, CA: NASA, pp. 210–216.

Kok, J. J., and Van Wijk, R. A. 1978. Evaluation of Models Describing Human Operator Control of Slowly Responding Complex Systems. Ph.D. thesis, Delft University of Technology, The Netherlands.

Krantz, D. H., Atkinson, R. D., Luce, R. D., and Suppes, P. (Eds.) 1974. *Contemporary Developments in Mathematical Psychology*. San Francisco: Freeman, Vols. I and II.

Kuhn, T. S. 1962. *The Structure of Scientific Revolutions*. Chicago: University of Chicago Press.

Landau, I. D. 1974. An Asymptotic Unbiased Recursive Identifier for Linear Systems. *Proceedings of the 1974 IEEE Decision and Control Conference*. Phoenix. New York: IEEE, pp. 288–294.

Lee, R. C. K. 1964. *Optimal Estimation, Identification, and Control*. Cambridge MA: MIT Press.

Licklider, J. C. R. 1960. Man–Computer Symbiosis. *IEEE Transactions on Human Factors in Electronics* HFE-1 (No. 1), 4–11.

Lilienfeld, R. 1978. *The Rise of Systems Theory: An Ideological Analysis*. New York: Wiley.

McRuer, D. T., Graham, D., Krendel, W. S., and Reisener, W., Jr. 1965. *Human Pilot Dynamics in Compensatory Systems—Theory Models and Experiments with Controlled Element and Forcing Function Variations*. Wright-Patterson AFB, OH: Air Force Flight Dynamics Laboratory, AFFDL-TR-65-15.

McRuer, D. T. and Krendel, E. S. 1957. *Dynamic Response of Human Operators*. Wright-Patterson AFB, OH: Wright Air Development Center, WADC-TR-56-524.

Meditch, J. S. 1969. *Stochastic Optimal Linear Estimation and Control*. New York: McGraw-Hill.

Mehra, R. K. 1970. On the Identification of Variances and Adaptive Kalman Filtering. *IEEE Transactions on Automatic Control* AC-15 (No. 2), 175–184.

Mehra, R. K. 1971. On-Line Identification of Linear Dynamic Systems with Applications to Kalman Filtering. *IEEE Transactions on Automatic Control* AC-16 (No. 1), 12–21.

Mendel, J. M. 1973. *Discrete Techniques of Parameter Estimation*. New York: Marcel Dekker.

Minsky, M.L. 1975. A Framework for Representing Knowledge. *In The Psychology of Computer Vision* (P. H. Winston, Ed.). New York: McGraw-Hill.

Moray, N. 1979. *Mental Workload*. New York: Plenum Press.

Morrison, N. 1969. *Introduction to Sequential Smoothing and Prediction*. New York: McGraw-Hill.

Muralidharan, R., and Baron, S. 1979. DEMON: A Human Operator Model for Decision Making, Monitoring and Control. *Proceedings of the Fifteenth Annual Conference on Manual Control*, Wright State University. Wright-Patterson AFB, OH: Air Force Flight Dynamics Laboratory.

Nelson, L. W., and Stear, E. 1976. The Simultaneous On-Line Estimation of Parameters and States in Linear Systems. *IEEE Transactions on Automatic Control* AC-21 (No. 1), 94–98.

Newell, A. 1973. Production Systems: Models of Control Structures. In *Visual Information Processing* (W. G. Chase, Ed.). New York: Academic Press, Chap. 10.

Newell, A., and Simon, H. A. 1972. *Human Problem Solving*. Englewood Cliffs, NJ: Prentice-Hall.

Pellegrino, S. J. 1979. Modeling Test Sequences Chosen by Humans in Fault Diagnosis Tasks. MSIE thesis, University of Illinois at Urbana-Champaign.

Penniman, W. D. 1975. A Stochastic Process Analysis of Online User Behavior. *Proceedings of the 38th Annual Meeting of the American Society for Information Science*, Boston. Washington, DC: American Society for Information Science, pp. 147–148.

Phatak, A. V. 1976. Formulation and Validation of Optimal and Control Theoretic Models of the Human Operator. *Man–Machine Systems Review* 2, 11–12.

Pritsker, A. A. B. 1974. *The GASP IV Simulation Language*. New York: Wiley.

Rasmussen, J. 1979. *On the Structure of Knowledge—A Morphology of Mental Models in a Man–Machine Context*. Roskilde, Denmark: Riso National Laboratory, RISO-M-2192.

Rasmussen, J., and Jensen, A. 1974. Mental Procedures in Real-Life Tasks: A Case Study of Electronic Trouble Shooting. *Ergonomics* 17 (No. 3), 293–307.

Rijnsdorp, J. E. and Rouse, W. B. 1977. Design of Man–Machine Interfaces in Process Control. In *Digital Computer Applications to Process Control* (H. R. Van Nauta Lemke and H. B. Verbruggen, Eds.). New York: North Holland.

Ross, S. M. 1970. *Applied Probability Models with Optimization Applications*. San Francisco: Holden-Day.

Rouse, W. B. 1972. Cognitive Sources of Suboptimal Human Prediction. Ph. D. thesis, MIT.

Rouse, W. B. 1973. A Model of the Human in a Cognitive Prediction Task. *IEEE Transactions on Systems, Man, and Cybernetics* SMC-3 (No. 5), 473–477.

Rouse, W. B. 1975. Design of Man–Computer Interfaces for On-Line Interactive Systems. *Proceedings of the IEEE* 63 (No. 6), 847–857.

Rouse, W. B. 1976. A Model of the Human as a Suboptimal Smoother. *IEEE Transactions on Systems, Man, and Cybernetics* SMC-6 (No. 5), 337–343.

Rouse, W. B. 1977a. A Theory of Human Decision Making in Stochastic Estimation Tasks. *IEEE Transactions on Systems, Man, and Cybernetics* SMC-7 (No. 4), 274–283.

Rouse, W. B. 1977b. Human–Computer Interaction in Multi-Task Situations. *IEEE Transactions on Systems, Man, and Cybernetics* SMC-7 (No. 5), 384–392.

Rouse, W. B. (Ed.) 1977c. Special Issue on Applications of Control Theory in Human Factors. *Human Factors* 19 (Nos. 4 and 5).

Rouse, W. B. 1978a. Human Problem Solving Performance in a Fault Diagnosis Task. *IEEE Transactions on Systems, Man, and Cybernetics* SMC-8 (No. 4), 258–271.

Rouse, W. B. 1978b. A Model of Human Decision Making in a Fault Diagnosis Task. *IEEE Transactions on Systems, Man, and Cybernetics* SMC-8 (No. 5), 357–361.

Rouse, W. B. 1979a. A Model of Human Decision Making in Fault Diagnosis Tasks That Include Feedback and Redundancy. *IEEE Transactions on Systems, Man, and Cybernetics* SMC-9 (No. 4), 237–241.

Rouse, W. B. 1979b. Problem Solving Performance of Maintenance Trainees in a Fault Diagnosis Task. *Human Factors* 21 (No. 2), 195–203.

Rouse, W. B. 1979c. Problem Solving Performance of First Semester Maintenance Trainees in Two Fault Diagnosis Tasks. *Human Factors* 21 (No. 5), 611–618.

Rouse, W. B., and Gopher, D. 1977. Estimation and Control Theory: Application to Modeling Human Behavior. *Human Factors* 19 (No. 4), 315–329.

Rouse, W. B., and Greenstein, J. S. 1976. A Model of Decision Making in Multi-Task Situations: Implications for Computer Aiding. *Proceedings of the 1976 International Conference on Cybernetics and Society*, Washington. New York: IEEE, pp. 425–433.

Rouse, W. B., and Rouse, S. H. 1979. Measures of Complexity of Fault Diagnosis Tasks. *IEEE Transactions on Systems, Man, and Cybernetics* SMC-9 (No. 11), 720–727.

Rouse, W. B., Rouse, S. H., and Pellegrino, S. J. 1979. A Rule-Based Model of Human Problem Solving Performance in Fault Diagnosis Tasks. University of Illinois at Urbana-Champaign.

Rouse, W. B., Rouse, S. H., and Pellegrino, S. J. 1980. A Rule-Based Model of Human Problem Solving Performance in Fault Diagnosis Tasks. *IEEE Transactions on Systems, Man, and Cybernetics* SMC-10.

Sacerdoti, E. D. 1975. A Structure for Plans and Behavior. Ph.D. thesis, Stanford University.

Sage, A. P., and Melsa, J. L. 1971. *System Identification*. New York: Academic Press.

Schank, R. C., and Abelson, R. P. 1977. *Scripts, Plans, Goals, and Understanding*. Hillsdale, NJ: Lawrence Erlbaum.

Schmidt, D. K. 1978. A Queueing Analysis of the Air Traffic Controller's Workload. *IEEE Transactions on Systems, Man, and Cybernetics* SMC-8 (No. 6), 492–493.

Schriber, T. J. 1974. *Simulation Using GPSS*. New York: Wiley.

Scott, A. C. 1977. *Neurophysics*, New York: Wiley.

Senders, J. W., and Posner, M. J. M. 1976. A Queueing Model of Monitoring and Supervisory Behavior. In *Monitoring Behavior and Supervisory Control* (T. B. Sheridan and G. Johannsen, Eds.). New York: Plenum Press.

Shannon, C. E., and Weaver, W. 1948. *The Mathematical Theory of Communication.* Urbana, IL: University of Illinois Press.

Sheridan, T. B. 1970. On How Often the Supervisor Should Sample. *IEEE Transactions on Systems Science and Cybernetics* SSC-6 (No. 2), 140–145.

Sheridan, T. B. 1976. Toward a General Model of Supervisory Control. In *Monitoring Behavior and Supervisory Control* (T. B. Sheridan and G. Johannsen, Eds.). New York: Plenum Press.

Sheridan, T. B., and Ferrell, W. R. 1974. *Man–Machine Systems: Information, Control, and Decision Models of Human Performance.* Cambridge, MA: MIT Press.

Sheridan, T. B., and Johannsen, G. (Eds.) 1976. *Monitoring Behavior and Supervisory Control.* New York: Plenum Press.

Simon, H. A. 1969. *The Sciences of the Artificial.* Cambridge, MA: MIT Press.

Sklansky, J. 1978. Image Segmentation and Feature Extraction. *IEEE Transactions on Systems, Man, and Cybernetics* SMC-8 (No. 4), 237–247.

Takahashi, Y., Rabins, M. J., and Auslander, D. M. 1970. *Control and Dynamic Systems.* Reading, MA: Addison-Wesley.

Tatsuoka, M. M. 1971. *Multivariate Analysis.* New York: Wiley.

Tomizuka, M. 1973. The Optimal Finite Preview Problem and Its Application to Man–Machine Systems. Ph.D. thesis, MIT.

Tomizuka, M., and Fujimura, M. 1979. Extended Signal Quickening for Manual Control. *IEEE Transactions on Systems, Man, and Cybernetics* SMC-9 (No. 10), 668–676.

Tomizuka, M., and Tam, W. M. 1978. An Extension of the Quickened Display for Manual Control. *Proceedings of the Fourteenth Annual Conference on Manual Control,* University of Southern California. Moffett Field, CA: NASA Conf. Publ. 2060, pp. 33–43.

Tomizuka, M., and Whitney, D. E. 1975. Optimal Finite Preview Problems (Why and How is Future Information Important?) *Journal of Dynamic Systems, Measurement, and Control* 97 (No. 4), 319–325.

Tomizuka, M., and Whitney, D. E. 1976. The Human Operator in Manual Preview Tracking (An Experiment and Its Modeling Via Optimal Control). *Journal of Dynamic Systems, Measurement, and Control* 98 (No. 4), 407–413.

Tong, R. M. 1977. A Control Engineering Review of Fuzzy Systems. *Automatica* 13 (No. 6), 559–569.

Walden, R. S. 1977. A Queueing Model of Pilot Decision Making in a Multi-Task Flight Management Situation. MSME thesis, University of Illinois at Urbana-Champaign.

Walden, R. S., and Rouse, W. B. 1978. A Queueing Model of Pilot Decision Making in a Multi-Task Flight Management Situation. *IEEE Transactions on Systems, Man, and Cybernetics* SMC-8 (No. 12), 867–875.

Warfield, J. N. 1976. *Societal Systems: Planning, Policy, and Complexity.* New York: Wiley.

Weisbrod, R. L., Davis, K. B., and Freedy, A. 1977. Adaptive Utility Assessment in Dynamic Decision Processes: An Experimental Evaluation of Decision Aiding. *IEEE Transactions on Systems, Man, and Cybernetics* SMC-7 (No. 5), 377–383.

Weissman, S. J. 1976. On a Computer System for Planning and Execution in Incompletely Specified Environments. Ph.D. thesis, Univeristy of Illinois at Urbana-Champaign.

Weizenbaum, J. 1976. *Computer Power and Human Reason.* San Francisco: Freeman.

Wesson, R. B. 1977. Planning in the World of the Air Traffic Controller. *Proceedings of the Fifth International Joint Conference on Artificial Intelligence,* MIT. Cambridge MA: MIT Artificial Intelligence Laboratory, pp. 473–479,

White, J. W., Schmidt, J. W., and Bennett, G. K. 1975. *Analysis of Queueing Systems.* New York: Academic Press.

Wiener, N. 1948. *Cybernetics.* Cambridge, MA: MIT Press.

Winston, P. H. (Ed.) 1975. *The Psychology of Computer Vision.* New York: McGraw-Hill.

Winston, P. H. 1977. *Artificial Intelligence.* Reading, MA: Addison-Wesley.

Zadeh, L. A. 1965. Fuzzy Sets. *Information and Control* 8, 338–353.

Zadeh, L. A. 1973. Outline of a New Approach to the Analysis of Complex Systems and Decision Processes. *IEEE Transactions on Systems, Man, and Cybernetics* SMC-3 (No. 1), 28–44.

Zadeh, L. A., Fu, K. S., Tanaka, K., and Shimura, M. (Eds.) 1975. *Fuzzy Sets and Their Applications to Cognitive and Decision Processes.* New York: Academic Press.

Ziman, J. 1968. *Public Knowledge: The Social Dimension of Science.* Cambridge: Cambridge University Press.

Author Index

Taylor, F. V., 53
Tomizuka, M., 56
Tong, R. M., 87, 94

V

Van Wijk, R. A., 57

W

Walden, R. S., 75, 83
Ward, J. L., 73
Warfield, J. N., 126

Weaver, W., 104
Weisbrod, R. L., 114
Weissman, S. J., 124
Weizenbaum, J., 103
Wesson, R. B., 105
White, J. W., 60, 67, 69
Whitney, D. E., 56
Wiener, N., 9
Winston, P. H., 103, 108, 118

Z

Zadeh, L. A., 87, 89
Ziman, J., 9

Subject Index